T0215583

This book provides a multidisciplinary introduction to the subject of Langmuir–Blodgett films. These films are the focus of intense current worldwide interest, as the ability to deposit organic films of nanometre thicknesses has many implications in materials science, and in the development of new electronic and opto-electronic devices.

Beginning with the application of simple thermodynamics to the common bulk phases of matter, the book outlines the nature of the phases associated with floating monolayer films. The Langmuir–Blodgett deposition process itself is described in some detail and contrasted with other thin film techniques. Monolayer-forming materials and the structural, electrical and optical properties of Langmuir–Blodgett films are discussed separately.

Each chapter is comprehensive, easy to understand and generously illustrated. Appendices are provided for the reader wishing to delve deeper into the physics and chemistry background.

Langmuir–Blodgett films

An introduction

Langmuir–Blodgett films

An introduction

MICHAEL C. PETTY
University of Durham, UK

CAMBRIDGE
UNIVERSITY PRESS

Published by the Press Syndicate of the University of Cambridge
The Pitt Building, Trumpington Street, Cambridge CB2 1RP
40 West 20th Street, New York, NY 10011-4211, USA
10 Stamford Road, Oakleigh, Melbourne 3166, Australia

First published 1996

A catalogue record for this book is available from the British Library

Library of Congress cataloguing in publication data
Petty, M. C.
Langmuir–Blodgett films: an introduction / M. C. Petty.
p.   cm.
Includes bibliographical references.
ISBN 0 521 41396 6. – ISBN 0 521 42450 X (pbk)
1. Thin layers, Multilayered.   I. Title.
QC176.9.M84P45   1996
530.4'275 – dc20. 95-2781 CIP

ISBN 0 521 41396 6   hardback
ISBN 0 521 42450 x   paperback

Transferred to digital printing 2004

for Anne

Contents

Preface

Langmuir–Blodgett (LB) films have been the subject of scientific curiosity for most of the twentieth century. However, interest has grown significantly since the 1970s – a direct result of the work of Hans Kuhn and colleagues on energy transfer in multilayer systems. This introduced the idea of *molecular engineering,* i.e., using the LB technique to position certain molecular groups at precise distances to others. In this way new thin film materials could be built up at the molecular level and the relationship between these artificial structures and the natural world explored.

There are already several books that cover LB and related thin films. So why another? My own background is in electronics. While I have been involved in LB film research I have spent many hours pondering on chemical formulae, struggling with biological nomenclature and trying to understand the finer points of thermodynamics. The scope of the subject is continuing to grow and anyone now starting work in the area must assimilate an enormous amount of information. My intention therefore has been to provide a gentle introduction to newcomers with an emphasis on the multidisciplinary and interdisciplinary nature of the field.

Each chapter addresses a different issue. Chapter 1 describes the various bulk phases of matter and outlines physical principles that can be used to model these. Monolayer phases are introduced in chapter 2. Much is still to be learnt about the nature of even the simplest monolayers and I have concentrated on experimental aspects rather than theory. The LB process itself is discussed in some detail in chapter 3. Again, my aim has been to provide useful practical information. I have also briefly reviewed related thin film techniques. Materials that can be manipulated at the air/water interface are described in chapter 4. In chapter 5 the main methods of investigation are introduced and some important results outlined. Chapter 6 covers the electrical behaviour. This is an area that has attracted considerable interest, not least because of the possibilities of exploiting LB layers in electronic device structures; I have included a section outlining such ideas. Finally, chapter 7 provides an overview of the optical properties of mono- and multi-layer films.

At the end of each chapter, I have given the key references. These should be good starting points for the reader wishing to delve deeper. Important subjects such as crystallography, chemical bonding and the interaction of electromagnetic radiation with thin films are dealt with in separate appendices.

I hope that the result is a book readily accessible to both new research students and more experienced workers.

Acknowledgements

I am indebted to colleagues at Durham and elsewhere for the encouragement to write this book. Members of the Durham Centre for Molecular Electronics have helped with many useful suggestions. I am particularly grateful to Jim Feast for the tutorials on organic chemistry and to John Cresswell for advice on nonlinear optics.

Original photographs were provided by Jas Pal Badyal and Simon Evenson, David Batchelder, Harald Fuchs and Mathias Lösche. Dirk Hönig spent many hours working on the cover photograph. Julie Morgan helped with the rest of the photography. Kay Seaton put in a tremendous effort drawing all the diagrams and deserves my very special thanks.

Notes for reference

Constants

velocity of light in free space c	$2.998 \times 10^8 \, \text{m s}^{-1}$
permittivity of free space, ϵ_0	$8.854 \times 10^{-12} \, \text{F m}^{-1}$
electronic charge, q	$1.602 \times 10^{-19} \, \text{C}$
Planck's constant, h	$6.626 \times 10^{-34} \, \text{J s}$
Boltzmann's constant, k	$1.381 \times 10^{-23} \, \text{J K}^{-1}$
Avogadro's number, N_A	$6.022 \times 10^{26} \, \text{kilomole}^{-1}$
universal gas constant, R	$8.314 \times 10^3 \, \text{J kilomole}^{-1} \, \text{K}^{-1}$

Useful relationships

1 electronvolt (eV) $= 1.602 \times 10^{-19} \, \text{J}$
for vacuum, energy in eV $= 1.240/(\text{wavelength in } \mu\text{m})$
$1 \, \text{eV} = 8066 \, \text{cm}^{-1}$
1 eV per particle $= 23\,060 \, \text{kcal kilomole}^{-1}$ ($=23.06 \, \text{kcal mole}^{-1}$)
1 calorie $= 4.186 \, \text{J}$
at 300 K, $kT \approx 1/40 \, \text{eV}$
1 atmosphere $= 1.013 \times 10^5 \, \text{N m}^{-2}$

The electromagnetic spectrum

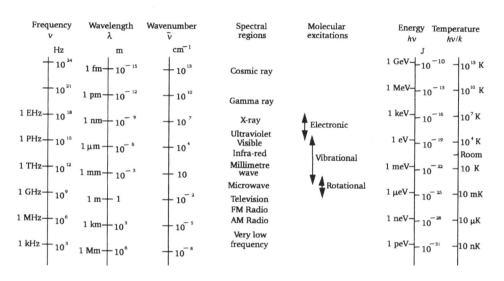

$$[1\text{Hz} \equiv 3.00 \times 10^8 \text{ m} \equiv 4.14 \times 10^{-15} \text{ eV} \equiv 4.80 \times 10^{-11} \text{ K}]$$

Simple functional groups

Compound type	Functional group
Alkene (double bond)	$\diagdown \text{C} = \text{C} \diagup$
Alkene (triple bond)	$-\text{C} \equiv \text{C}-$
Alkene (aromatic ring)	(benzene ring structure)
Halide	$-\text{C}-\text{X}$ (X - F, Cl, Br, I)
Alcohol	$-\text{C}-\text{O}-\text{H}$
Ether	$-\text{C}-\text{O}-\text{C}-$
Carbonyl	$\diagdown \text{C} = \text{O}$
Aldehyde	$\text{H}-\diagdown \text{C} = \text{O}$, $-\text{C}$
Ketone	$-\text{C}$, $\diagdown \text{C} = \text{O}$, $-\text{C}$
Ester	$-\overset{\text{O}}{\overset{\|}{\text{C}}}-\text{O}-\text{C}-$

Compound type	Functional group
Carboxylic acid	$-\overset{\text{O}}{\overset{\|}{\text{C}}}-\text{O}-\text{H}$
Carboxylic acid chloride	$-\overset{\text{O}}{\overset{\|}{\text{C}}}-\text{Cl}$
Amide	$-\overset{\text{O}}{\overset{\|}{\text{C}}}-\text{N} \diagup$
Amine	$-\text{C}-\text{N} \diagup$
Nitrile	$-\text{C} \equiv \text{N}$
Nitro	$-\text{C}-\overset{+}{\text{N}} \diagup \overset{\diagup \text{O}}{\diagdown \text{O}^-}$
Sulfide	$-\text{C}-\text{S}-\text{C}-$
Sulfoxide	$-\text{C}-\overset{\text{O}^-}{\underset{+}{\text{S}}}-\text{C}-$
Sulfone	$-\text{C}-\overset{\text{O}^-}{\underset{\text{O}^-}{\overset{2+}{\text{S}}}}-\text{C}-$
Organometallic	$-\text{C}-\text{M}$ (M - metal)

1

The bulk phases of matter

1.1   Gases, liquids and solids

The three most common states, or *phases*, of matter, *gases*, *liquids* and *solids* are very familiar (Walton, 1976). Phases that are not so well known are *plasmas* and *liquid crystals* (although these are both found in electrical and electronic devices in everyday use). All these states are generally distinguished by the degree of *translational* and *orientational order* of the constituent molecules. On this basis some phases may be further subdivided. For example, solids, consisting of a rigid arrangement of molecules, can be *crystalline* or *amorphous*. In an amorphous solid (a good example is a glass), the molecules are fixed in place, but with no pattern in their arrangement. As shown in figure 1.1, the crystalline solid state is characterized by long-range translational order of the constituent molecules (the molecules are constrained to occupy specific positions in space) and long-range orientational order (the molecules orient themselves with respect to each other). The molecules are, of course, in a constant state of thermal agitation, with a mean translational kinetic energy of $3kT/2$ (k is Boltzmann's constant, T is temperature; $kT/2$ for each component of their velocity). However, this energy is considerably less than that associated with the chemical bonds in the material and the motion does not disrupt the highly ordered molecular arrangement.

In the gaseous state, the intermolecular forces are not strong enough to hold the molecules together. They are therefore free to diffuse about randomly, spreading evenly throughout any container they occupy, no matter how large it is. The average interatomic distance is determined by the number of molecules and the size of the container. A gas is easily compressed, as it takes comparatively little force to move the molecules closer together.

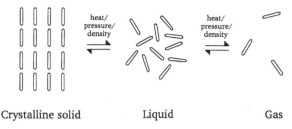

Crystalline solid   Liquid   Gas

Figure 1.1 The three most common bulk phases of matter.

On the microscopic level, the liquid state is generally thought of as a phase that is somewhere between that of a solid and that of a gas. The molecules in a liquid neither occupy a specific average position nor remain oriented in a particular way. They are free to move around and, as in the gaseous state, this motion is random. The physical properties of both liquids and gases are *isotropic*; i.e., they do not depend upon direction (direction-dependent properties are called *anisotropic*). Therefore, there is a similarity between liquids and gases and under certain conditions it is impossible to distinguish between these two states. When placed in a container, the liquid will fill it to the level of a free surface. Liquids flow and change their shape in response to weak outside forces. The forces holding a liquid together are much less than those in a solid. Liquids are highly incompressible, a characteristic that is exploited in hydraulic systems.

Almost all elements and chemical compounds possess a solid, a liquid and a gaseous phase. The gaseous phase is favoured by high temperatures and low pressures. A transition from one phase to another can be provoked by a change in temperature, pressure, density or volume.

1.2 Liquid crystals

The melting of a solid is normally a very sharp transition. As the material is heated, the order of the system abruptly changes from that of the three-dimensional order of the solid to the zero order associated with the liquid. However, it is not uncommon for an *organic* solid to pass through inter-mediate phases as it is heated from a solid to a liquid. Perhaps one in every few hundred organic compounds exhibits such behaviour. Such phases are known as *mesophases*. When such a state is formed, the translational order of the solid phase may be lost, but the orientational order remains. The result-ing phase is termed a *thermotropic liquid crystal* (Collings, 1990). Sometimes a mesophase may display translational order but no orientational order; this is termed a *plastic crystal*. A very wide range of liquid crystal phases is now known. These are identified by the degree of short range translational order and by the shape of the molecules.

Director

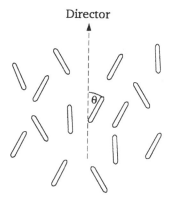

Figure 1.2 *Arrangement of rod-shaped molecules in a liquid crystalline phase. Each molecule makes an angle of θ with the director.*

The *nematic* (from the Greek word for thread) phase is the least ordered liquid crystal phase and is exploited extensively in electrooptic applications. This phase has no long-range translational order and only orientational order. There is only one nematic phase and, on heating, this will eventually become an isotropic liquid. A schematic diagram, showing the arrangement of rod-shaped molecules in a nematic phase, is given in figure 1.2. The molecules in the liquid crystal are free to move about in much the same fashion as in a liquid; as they do so they tend to remain orientated in a certain direction. The direction of preferred orientation is called the *director* of the liquid crystal. Each molecule is oriented at some angle θ about the director. The degree of orientational order is given by the *order parameter S* which is defined

$$S = \tfrac{1}{2} \langle 3 \cos^2 \theta - 1 \rangle \qquad (1.1)$$

A value of $S = 1$ indicates perfect orientational order whereas no orientational order results in $S = 0$.

Smectic (from the Greek word for soap) phases are usually formed by thermotropic liquid crystals at lower temperature than the nematic phase. Besides the orientational order, smectic phases possess one-dimensional translational ordering into layers. The smectic phases can be further sub-divided and, at present, twelve different types have been identified. These are designated $S_A, S_B, \ldots S_K$; there are two smectic S_B phases – the *crystal B* and *hexatic* (for a discussion of this phase, see Tredgold 1994) S_B phases. Some of these mesophases (crystal B, S_E, S_G, S_H and S_J) have very long-range correlation of position over many layers and are more similar to crystalline solids.

The S_A phase is the least ordered of the thermotropic smectic phases. The molecules are arranged in disordered layers, each layer having a liquid-like

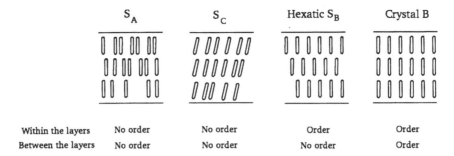

	S_A	S_C	Hexatic S_B	Crystal B
Within the layers	No order	No order	Order	Order
Between the layers	No order	No order	No order	Order

Figure 1.3 Different smectic liquid crystalline arrangements of rod-shaped molecules. S_C is a tilted version of S_A.

freedom of motion of its constituent molecules in two dimensions, with the director perpendicular to the layer planes. By contrast, in the S_C phase the molecules are tilted from this direction by about 35°. This tilt is correlated between molecules within each layer and from one layer to another. In the hexatic S_B phase, there is again ordering of orientationally aligned molecules into layers. The molecules are arranged in an hexagonal array but the translational order is short-range only. The orientation of the hexagonal net is, however, maintained over a long range and unlike the ordering of molecular positions, is correlated between layers. Two tilted variations of the S_B phase exist in which the tilt direction is constrained to point either towards one face of the hexagonal lattice (S_F) or towards one apex (S_I). The S_A, S_C, hexatic S_B and crystal B arrangements of rod-shaped molecules are contrasted in figure 1.3 (Lacey, 1994). Further details about these smectic liquid crystal phases are to be found in the book by Gray and Goodby (1984).

The final distinct type of liquid crystalline mesophase is the *cholesteric* or *chiral nematic*. The molecules in such a phase are arranged in a unique helical structure in which the director gradually rotates from one plane of molecules to the next. Such phases can exhibit vivid colour effects that change with temperature (as the pitch of the helix changes) and are exploited as temperature sensors. These molecules may also form smectic phases.

1.3   Phase changes

The science of thermodynamics allows the properties of the various phases of matter to be correlated. The *state* of any thermodynamic system (e.g., a fixed amount of matter) is specified by values of certain experimentally variable quantities called *state variables* or *properties*. Familiar examples are the temperature of a system, its pressure and the volume occupied by it. Those properties that are proportional to the mass of the system are called *extensive*; examples are the total volume and total energy of the

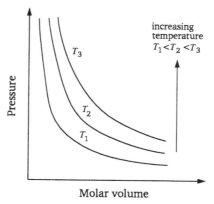

Figure 1.4 Pressure versus molar volume isotherms for an ideal gas.

system. Properties such as temperature or density, which are independent of mass, are called *intensive*.

It is found by experiment that only a certain minimum number of the properties of a pure substance can be given an arbitrary value. The values of the remaining properties are then determined by the nature of the substance. For example, suppose that nitrogen gas is allowed to flow into an evacuated tank at a constant temperature T. The volume V of the gas admitted is then fixed by the volume of the tank and the number of moles n of gas is fixed by the amount that is admitted. Once T, V and n have been fixed, the pressure P is determined by the nature of nitrogen and cannot be given any arbitrary value. It follows that there exists a certain relation between the properties P, V, T and n. This relation is termed an *equation of state* and, for an ideal gas, may be written

$$Pv = RT = N_A kT \tag{1.2}$$

where R is the *universal gas constant* ($R = 8.314 \times 10^3 \, \mathrm{J\, kilomole^{-1}\, K^{-1}}$), N_A is Avogadro's number ($N_A = 6.022 \times 10^{26} \, \mathrm{kilomole^{-1}}$) and v is the *specific molar volume* (i.e., $v = V/n$). Boltzmann's constant, k ($k = 1.381 \times 10^{-23} \, \mathrm{J\, K^{-1}}$) may be thought of as the *universal gas constant per molecule* (i.e., $k = R/N_A$).

Figure 1.4 shows a series of pressure versus volume plots for an ideal gas at different temperatures. The curves are simple hyperbolas and are called *isotherms*. It is also convenient to plot the values of P, v and T to form a three-dimensional graph. For a real substance, such a representation will include data over the liquid and solid phase regions. Such a diagram, for a substance that contracts on freezing (e.g., carbon dioxide), is shown in figure 1.5; this is known as a *phase diagram*. The figure reveals that there are certain regions (i.e., certain ranges of the variables P, v and T) in which

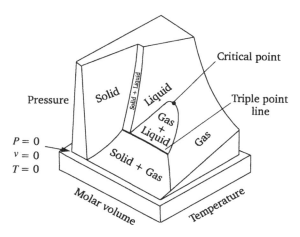

Figure 1.5 Pressure versus molar volume versus temperature plot for a substance that contracts on freezing.

the substance can exist in a single phase only; these are labelled solid, liquid and gas. In other regions, labelled solid + gas, solid + liquid and gas + liquid, two phases can exist simultaneously. All three phases exist along a line known as the *triple point line*. At a particular point called the *critical point*, the specific molar volumes of the liquid and gas become equal and it is no longer possible to distinguish one of these phases from the other.

Equations of state may also be written for both the liquid and the solid regions of the phase diagram shown in figure 1.5. For example, the following expression, derived by van der Waals, may apply to both the liquid and gaseous regions

$$\left(P + \frac{a}{v^2}\right)(v - b) = RT \tag{1.3}$$

a and b are constants for any one gas/liquid: the term a/v^2 arises from the existence of intermolecular forces and the constant b is proportional to the volume occupied by the molecules themselves. In the case of a solid, a simple equation of state might be

$$v = l + mT + nP \tag{1.4}$$

where l, m and n are empirically determined constants.

Isotherms for a real substance can be obtained by projecting the v–P–T surface onto the pressure–volume plane. Figure 1.6 is such a construction, illustrating what occurs as the specific molar volume of a material is reduced, taking it from the gaseous state to its solid phase. Curves are shown at three different temperatures. At temperature T_2, the first stage of the process

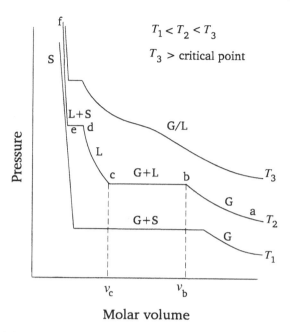

$T_1 < T_2 < T_3$

$T_3 >$ critical point

Figure 1.6 Pressure versus molar volume isotherms at three different temperatures for a bulk material.

(a → b in the figure) is the isothermal compression of the gas. If the gas is ideal, the P versus v curve will be a simple hyperbola, as shown in figure 1.4. At point b the gas begins to condense to a liquid. The pressure remains constant until point c is reached, when all the gas has condensed. Between the pure gas state at b and the pure liquid state at c, the ratio of the mass of gas present (G) to the total mass of gas and liquid (G + L) decreases linearly from one to zero. The proportion by mass should not be confused with the proportion by volume. Throughout the change from b to c, the molar volume of the gas is v_b and that of the liquid is v_c. Further reduction in volume of the liquid phase causes the appearance of solid particles at point d. Solidification (at constant pressure) is complete at point e at which only single phase solid is present. Curve e → f therefore represents the compression of the solid phase. Note that if the compression takes place at a temperature above the critical point (i.e., at T_3 in figure 1.6), the distinct phase change corresponding to the condensation of the gas to the liquid is not observed; the gas and liquid phases are now inseparable. On the other hand, if the temperature is reduced sufficiently (i.e., to T_1), then the gas transforms directly into a solid.

Phase changes, such as solidification or vaporization, which take place at constant temperature, are accompanied by heat absorption or evolution.

This heat is commonly known as the *latent heat* of the transformation. However, in the strict thermodynamic sense, this is the *enthalpy H* given by

$$H = U + PV \qquad (1.5)$$

where the first term, U, on the right-hand side of the equation is the *internal energy*. This is simply the sum of all the potential and kinetic energies in a system. The above relation shows that the enthalpy associated with any phase change consists of two distinct parts. For example, when a liquid vaporizes at its boiling point, energy is required not only to break the bonds between molecules in the liquid (the U term) but also to 'push back' the atmosphere to make room for the vapour (the PV term). The horizontal (constant pressure) lines on the isotherms in figure 1.6 are associated with enthalpy changes as the phase transforms from gas to liquid ($b \rightarrow c$) or from liquid to solid ($d \rightarrow e$). At the critical temperature, the enthalpy for vaporization becomes zero.

1.4 Thermodynamic equilibrium

When an arbitrary system is left to itself, its properties (P, V, T etc.) will generally change. However, after a sufficiently long period, the properties will be invariant with time. This final state will depend on the nature of the system. For example, a simple mechanical arrangement, such as a ball rolling down a hill, consistently comes to rest in the state of lowest energy (i.e., with the ball at the bottom of the hill). Any thermodynamic system, left for long enough, will reach a state of *thermodynamic equilibrium*. Here, thermodynamic equilibrium is achieved when the *free energy* is minimized. For systems at constant volume, this is given by the *Helmholtz function F*

$$F = U - TS \qquad (1.6)$$

and for constant pressure changes, the free energy is given by the *Gibbs' function G*

$$G = H - TS = U + PV - TS \qquad (1.7)$$

In both the above equations, S is the *entropy*. Classically this is the heat into, or out of, the system divided by temperature. For a simple phase change

$$\Delta S = \frac{\Delta H}{T} \qquad (1.8)$$

Entropy is also a measure of the disorder of a thermodynamic system. The specific molar entropy change ($\Delta S/n$) for vaporization is normally much greater than that for fusion, confirming the increased disorder produced in the former phase transformation.

Equations (1.6) and (1.7) reveal that free energy can be minimized either by reducing the internal energy or by increasing the entropy term. At low temperature, the internal energy of the molecules makes the greater contribution to the free energy, and so the solid phase is the favoured state. At higher temperature the entropy of the system becomes the predominant influence. As a result, the fluid phases are stable at elevated temperatures although they constitute higher internal energy configurations than the solid phase.

Many states of matter that are commonly encountered are not in a true state of thermodynamic equilibrium. Equilibrium may be approached very slowly (e.g., an amorphous glass will eventually crystallize over a period of many hundreds of years) or the system may be in a *metastable* state (e.g., a supercooled liquid or vapour). However, as such systems will have directly measurable properties (*P*, *H*, *S* etc.) that are stable during an experiment, it is appropriate to assume these properties can be related in the same way as for a true equilibrium state. A discussion on this topic may be found in many textbooks (e.g., Sears and Salinger, 1975).

Thermodynamically, phase changes of matter can be divided into *first-order* and *higher-order transitions*. In the former case, the specific molar Gibbs' function $g (= G/n)$ is continuous but there is a discontinuity in the first derivative of g across the transition. An example of such a change is that of melting or evaporation: in these cases the discontinuity in $\partial g/\partial T$ is simply equal to the entropy of transformation ($\Delta H/T$). For second-order thermodynamic transitions, both the specific Gibbs' function and its first derivative are continuous, but the second derivative changes discontinuously. In such transitions, the enthalpy of transformation is zero and the molar specific volume does not change. The transition of a liquid to a vapour at the critical point, ferromagnetic to paramagnetic transitions and the change of a superconductor from the superconducting state to the normal state in zero magnetic field are examples of second-order phase transitions.

1.5 Phase rule

A very useful tool for predicting equilibrium-phase relationships is the *Gibbs' phase rule* (Ferguson and Jones, 1973; Sears and Salinger, 1975). This can be written

$$F = C - P + 2 \tag{1.9}$$

where P is the number of phases and F is the number of *degrees of freedom* the system possesses. The latter quantity is simply the number of state variables (*T*, *V*, *P* etc.) that can be *independently* changed. The number of *components* (*C*) can be thought of as the number of 'pure' substances in the system. A formal definition (and precise definitions are extremely important in thermodynamics) is that the number of components is the minimum number of molecular species in terms of which the compositions of all the phases may be quantitatively expressed (Ferguson and Jones, 1973). Equation (1.9) relates to all macroscopic systems influenced only by changes in pressure, temperature and concentration.

The application of the phase rule may be illustrated by reference to the isotherm at temperature T_2 in figure 1.6. Here a single pure material ($C = 1$) is held at a constant temperature. In the gaseous phase region ($P = 1$), the pressure or the volume of the gas may be changed. These are not independent quantities as they are related by an equation such as (1.2); if the pressure is changed, then the volume of the gas will also change. The temperature, of course, is an additional degree of freedom. Therefore, for the single component, single phase system, $F = 2$, as predicted by equation (1.9). As the gas is compressed, the liquid state will appear; i.e., $P = 2$. The phase rule indicates that the thermodynamic system must lose one of its degrees of freedom ($F = 1$). The temperature has already been fixed; the pressure must therefore remain constant during the gas to liquid phase transformation. The pressure (or temperature) can vary again when all the gas has condensed into the liquid state. A similar argument may be used to explain the constant pressure region in the liquid to solid phase change in figure 1.6. If a single component system is held at constant pressure, rather than at a fixed temperature, then the phase rule may also be used to account for the familiar observation of a constant temperature during a gas to liquid or liquid to solid transformation.

If three phases of a single component system are in equilibrium, then the system has no degrees of freedom. This occurs along the triple point line in the full phase diagram shown in figure 1.5. At first sight, it might be thought that such a system has a single degree of freedom (the molar volume) as the gas/liquid/solid is in equilibrium along a line in this three-dimensional plot. However, it is only the relative *composition* of the phases that change along this line; the specific molar volumes of the three phases are fixed. The temperature and pressure of the triple point of a pure substance are uniquely defined and this is often used as a fixed point on a thermodynamic temperature scale.

The ideas developed in this chapter have concerned the relationships between the bulk phases of matter. However, many of the concepts are quite general and, as will be shown in chapter 2, may be equally well applied to the states of a monomolecular film.

References

Collings, P. J. (1990) *Liquid Crystals*, Adam Hilger, Bristol
Ferguson, F. D. and Jones, T. K. (1973) *The Phase Rule*, 2nd edition, Butterworths, London

Gray, G. W. and Goodby, J. W. G. (1984) *Smectic Liquid Crystals*, Leonard Hill, Glasgow

Lacey, D. (1995) Liquid crystals and devices, in *An Introduction to Molecular Electronics*, eds. M. C. Petty, M. R. Bryce and D. Bloor, Edward Arnold, London

Sears, F. W. and Salinger, G. L. (1975) *Thermodynamics, Kinetic Theory and Statistical Thermodynamics*, Addison-Wesley, Reading, MA

Tredgold, R. H. (1994) *Order in Thin Organic Films*, Cambridge University Press, Cambridge

Walton, A. J. (1976) *Three Phases of Matter*, McGraw-Hill, Maidenhead

2

Monolayers:
two-dimensional phases

2.1 The gas–liquid interface

Certain organic molecules will orient themselves at the interface between a gaseous and a liquid phase (or between two liquid phases) to minimize their free energy. The resulting surface film is one molecule in thickness and is commonly called a *monomolecular layer* or simply a *monolayer*. In the previous chapter the individual properties of bulk phases were outlined. The interface region will now be examined.

The boundary between a liquid and a gas (e.g., the air/water interface) marks a transition between the composition and properties of the two bulk phases. A surface layer will exist with different properties from those of either bulk phase (Adamson, 1982; Gaines, 1966). The thickness of this region is very important. If the molecules are electrically neutral, then the forces between them will be short-range and the surface layer will be no more than one or two molecular diameters. In contrast, the Coulombic forces associated with charged species can extend the transition region over considerable distances.

The microscopic model of a real interface is one of dynamic molecular motion as molecules move in and out of it. However, for the interface to be in equilibrium, as many molecules must diffuse from the bulk of the liquid to its surface per unit time as leave the surface for the bulk. A molecule at the surface is surrounded by fewer molecules than one in the bulk liquid (see figure 2.1). Therefore, more molecules will diffuse initially from the surface, increasing the mean atomic separation (and therefore the inter-molecular forces) between surface molecules. The activation energy for a surface molecule escaping into the bulk will then increase until it is equal to

12

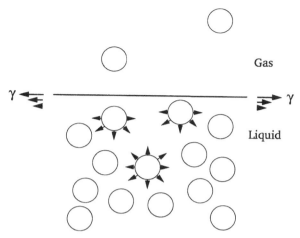

Figure 2.1 Forces experienced by molecules in the bulk of a liquid and at the liquid/gas interface.

that for molecules diffusing from the bulk to the surface, and a state of equilibrium is achieved. The line force acting on the surface molecules is the *surface tension* γ (Walton, 1976).

At thermodynamic equilibrium, the surface tension of a planar interface, can be shown to be related to the partial derivatives of the free energy functions, discussed in chapter 1, with respect to the area A of the surface (Birdi, 1989), i.e.,

$$\gamma = \left(\frac{\partial F}{\partial A}\right)_{T,V,n_i} = \left(\frac{\partial G}{\partial A}\right)_{T,P,n_i} \tag{2.1}$$

When describing the thermodynamics of surfaces, it is common practice to introduce the concept of *excess* quantities (Birdi, 1989; Gaines, 1966). Therefore, the thermodynamic extensive state variables of the two bulk phases (e.g., number of moles, internal energy, entropy) are assumed to be constant up to an imaginary dividing surface and the excess quantities describe the properties of the surface. For a *pure* liquid in equilibrium with its saturated vapour at a planar interface, surface tension is equal to the excess Helmholtz free energy per unit area, i.e.,

$$\gamma = \frac{F^S}{A} \tag{2.2}$$

where F^s refers to the surface excess free energy. There is also an alternative approach to the thermodynamics of surfaces in which the surface region is considered as a separate phase (of small but finite volume) with its own values of extensive variables (Gaines, 1966).

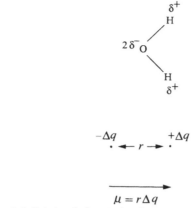

Figure 2.2 Origin of electric dipole moment, μ, in a water molecule.

Surface tension is analogous to vapour pressure, remaining constant for two phases in equilibrium if the temperature is constant, but changing with changing temperature. Unlike the vapour pressure, however, which increases with increasing temperature, γ decreases with increasing temperature and becomes zero at the critical point.

The presence of a monomolecular film on a liquid surface will affect the surface tension. In monolayer experiments it is normal to refer to a measurement of *surface pressure*. This pressure Π is equal to the reduction of the pure liquid surface tension by the film, i.e.,

$$\Pi = \gamma_0 - \gamma \qquad (2.3)$$

where γ_0 is the surface tension of the pure liquid and γ is the surface tension of the film-covered surface. Monomolecular films have been studied on a variety of *subphase* (liquid) surfaces, for example, water, mercury and hydrocarbons. However, almost all the work on transferred monolayers (i.e., removed from the subphase surface and deposited onto solid supports) has concentrated on aqueous subphases. The discussion will therefore be restricted to this. Values of Π of the order of mN m^{-1} are generally encountered in monolayer studies on a water surface; the maximum value of Π is 72.8 mN m^{-1} at 20 °C, the surface tension of water.

2.2 Monolayer materials

All compounds may be roughly divided into those that are soluble in water and those that do not dissolve. The former materials are generally *polar*, that is they carry an uneven distribution of charge. Associated with such molecules is an entity called a *dipole moment μ*. Figure 2.2 illustrates the case for a molecule of water. Each of the two hydrogen atoms shares an

Table 2.1 *Dipole moments of various chemical groups.*

Chemical group	Example	Dipole moment, μ [Debye]
OOH	propranoic acid (C_2H_5COOH)	1.75
OH	propanol (C_3H_7OH)	1.68
NO_2	nitropropane ($C_3H_7NO_2$)	3.66
NH_2	propylamine ($C_3H_5NH_2$)	1.17
CN	cyanoethane (C_2H_5CN)	4.02
benzene ring	(C_6H_6)	0
cis C=C	*cis*-2-butene	0.33
trans C=C	*trans*-2-butene	0

electron pair with the oxygen atom. The geometry of the shared electron pairs in the outer shell of the oxygen causes the molecule to be V-shaped. The strong electron-withdrawing tendency of the oxygen atom results in a local negative charge at the apex of the V, and gives the two hydrogen nuclei local positive charges. Although the water molecule is electrically neutral, its positive and negative charges are widely separated. If the water molecule is modelled as two equal and opposite charges $\pm\Delta q$ separated by a distance of r (constituting a dipole), then the magnitude of the dipole moment is given by the product of Δq and r. To be strictly correct, μ is a vector quantity, directed from negative to positive charge. The SI units of dipole moment are [C m], however most workers use the *Debye unit* ($1D = 3.336 \times 10^{-30}$ C m); the dipole moment of water is 1.85 D. The solvent properties of water are related to the attraction between its electric dipoles and the charges associated with the solute. Dipole moments associated with chemical groups that are commonly encountered in monolayer work are given in table 2.1. Symmetrical molecules, such as benzene, possess a zero dipole moment. These substances do not readily dissolve in water and are termed *nonpolar*.

The molecules of most (but certainly not all) monolayer-forming materials are composed of two parts: one that by itself would mix with water and another that by itself would not. The soluble part is called *hydrophilic* (water-loving) and the insoluble part is termed *hydrophobic* (water-hating). These molecules are *amphiphiles*, the most important types of which are *soaps* and *phospholipids*. Such compounds are also called *surfactants*. A full discussion of the range of these compounds is given in chapter 4. However, at this point, it is appropriate to introduce the simplest substances of this type. Figure 2.3 shows the structure of a classical monolayer forming material, n-octadecanoic acid, commonly known as stearic acid – $C_{17}H_{35}COOH$. This is one of a series of long-chain fatty acids, with the general chemical formula $C_nH_{2n+1}COOH$. The n-octadecanoic acid molecule

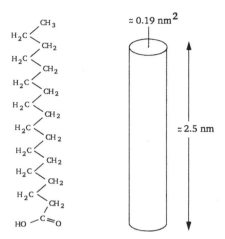

Figure 2.3 Chemical formula for n-octadecanoic acid (stearic acid). The approximate geometrical shape and dimensions of the molecule are shown on the right.

is cylindrical in shape, with a length of approximately 2.5 nm and a cross-section of about $0.19 \, nm^2$. It consists of 16 CH_2 groups forming a long hydrocarbon chain with a methyl CH_3 group at one end and a polar carboxylic acid COOH group at the other. The latter group confers water solubility while the hydrocarbon chain prevents it. It is the balance between these two forces that provides the monolayer forming abilities of the n-octadecanoic acid. If the hydrocarbon chain were too short, or the polar group too strong, then the material would simply dissolve in the subphase.

2.3   Monolayer stability

Most monolayer materials are applied to the subphase surface by first dissolving them in a suitable solvent (e.g., chloroform). When the solvent has evaporated, the organic molecules may be compressed to form a floating 'two-dimensional' solid. The hydrophilic and hydrophobic ends of the molecules ensure that, during this process, the individual molecules are aligned in the same way.

When the material is first applied to the surface of water, spreading will continue until the surface pressure has risen to an equilibrium value. This *equilibrium spreading pressure* is defined as that spontaneously generated when the bulk solid is placed in contact with a water surface (Iwahashi et al., 1985). Again, a comparison may be drawn with the vapour pressure of a bulk solid. An equilibrium vapour pressure exists for the solid in the presence of its vapour. If this vapour pressure is exceeded, i.e., the vapour becomes *supersaturated,* deposition onto the solid surface will occur. This should also be the case for a monolayer, which is expected to form crystals if its surface pressure is greater than the equilibrium spreading pressure. The formation of small droplets or lenses is often observed when excess monolayer material is spread. However, in many experiments, seemingly stable surface pressures up to higher values than the equilibrium spreading pressure may be measured. Deposition from such compressed layers may be compared to nucleation from supercooled or supersaturated bulk states.

The solubility of n-octadecanoic acid in water at 20 °C is approximately 290 μg per 100 ml. In a typical monolayer experiment, a few tens of μg of the compound might be spread on the surface of several hundred ml of water. The resulting compressed monolayer film is stable (if the surface pressure is not too high) over a long period. There is a barrier to dissolution and equilibrium may be approached only very slowly. Consequently, *a floating monomolecular film may be in a metastable state rather than in true thermodynamic equilibrium.*

2.4 Monolayer phases

As the monolayer is compressed on the water surface, it will undergo several phase transformations. These are, almost, analogous to the three-dimensional gases, liquids and solids described in chapter 1. The phase changes may be readily identified by monitoring the surface pressure Π as a function of the area occupied by the film. This is the two-dimensional equivalent of the pressure versus volume isotherm for a gas/liquid/solid (see figure 1.6). Figure 2.4 shows such a plot for a hypothetical long chain organic monolayer material (e.g., a long-chain fatty acid). This diagram is not meant to represent that observed for any particular substance, but shows most of the features observed for long chain compounds (Gaines, 1966).

In such a plot, it is usual to divide the film area A by the total number of molecules on the water surface to obtain the *area per molecule, a,* i.e.

$$a = \frac{AM}{CN_AV} = \frac{A}{cN_AV} \tag{2.4}$$

where M is the molecular weight of the monolayer material, C is the concentration of the spreading solution in mass per unit volume, c is the specific molar concentration of the solution and V is its volume.

In the 'gaseous' state (G in figure 2.4), the molecules are far enough apart on the water surface that they exert little force on one another. As the surface area of the monolayer is reduced, the hydrocarbon chains will begin to interact. The 'liquid' state that is formed is generally called the *expanded* monolayer phase (E). The hydrocarbon chains of the molecules in such a

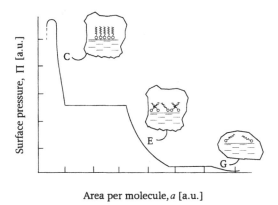

Figure 2.4 Surface pressure versus area per molecule isotherm for a long-chain organic compound. (The surface pressure and area are in arbitrary units, [a.u.].)

film are in a random, rather than a regular orientation, with their polar groups in contact with the subphase. As the molecular area is progressively reduced, *condensed* (C) phases may appear. There may be more than one of these and the emergence of each condensed phase can be accompanied by constant pressure regions of the isotherm, as observed in the cases of a gas condensing to a liquid and a liquid solidifying (cf. figure 1.6). These regions will be associated with enthalpy changes in the monolayer. In the condensed monolayer states, the molecules are closely packed and are oriented with the hydrocarbon chain pointing away from the water surface. The area per molecule in such a state will be similar to the cross-sectional area of the hydrocarbon chain, i.e., $\approx 0.19\,\text{nm}^2\,\text{molecule}^{-1}$.

It is found that monolayers can be compressed to pressures considerably higher than the equilibrium spreading pressure. The surface pressure continues to increase with decreasing surface area until a point is reached where it is not possible to increase the pressure any further and the area of the film decreases if the pressure is kept constant, or the pressure falls if the film is held at constant area. This is referred to as *collapse*. The forces acting on a monolayer at this point are quite high. For example, a surface pressure of $100\,\text{mN m}^{-1}$ acting on a layer of molecules 2.5 nm high corresponds to a three-dimensional pressure of about 400 atmospheres. The onset of collapse depends on many factors, including the rate at which the monolayer is being compressed and the history of the film. Therefore, to be meaningful, the pressure value should specify the exact conditions for collapse. When collapse occurs, molecules are forced out of the monolayer, as illustrated in figure 2.5.

Other structures are also possible for amphiphilic molecules. For example, if the molecules possess a strong polar head group relative to the nonpolar part

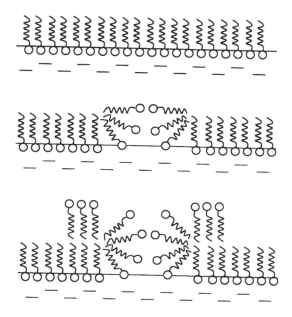

Figure 2.5 Stages of collapse of a monolayer, going from the top to the bottom diagram.

of the molecule, *micelles* are formed if the concentration of the amphiphilic material is above a certain value (*critical micelle concentration*) (Harkins, 1952). In this arrangement, the molecules are arranged in spheres, with the polar head groups on the outside and the hydrocarbon chains towards the centre. If the head group of the amphiphilic molecule is not strong compared with the hydrophobic part, the molecules can form spherical *vesicles* in which the double layers form a shell with water both outside and inside. Such structures are very common in the biological world with long-chain molecules that possess two hydrocarbon chains for each polar head group (phospholipids). Schematic diagrams of both micelles and vesicles are shown in figure 2.6.

2.4.1 Gaseous phase

If the behaviour of gaseous monolayers can be modelled using a two-dimensional variation of conventional kinetic theory, then the molecules in the film may be assumed to move about with an average translational kinetic energy of $kT/2$ for each degree of freedom. This leads to the following 'ideal' gaseous monolayer equation (cf. equation (1.2) for a perfect three-dimensional gas)

$$\Pi a = kT \qquad (2.5)$$

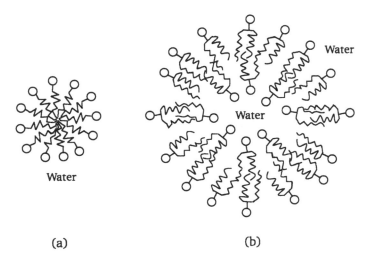

Figure 2.6 Cross sections of (a) a micelle and (b) a vesicle formed by amphiphilic molecules.

If Π is measured in [$\mathrm{mN\,m^{-1}}$] and a is in [$\mathrm{nm^2\ molecule^{-1}}$], their product should be 3.97 at 15 °C. Therefore, the surface pressure in the gaseous phase would be expected to be about $1\,\mathrm{mN\,m^{-1}}$ at an area per molecule of $4\,\mathrm{nm^2}$. This is consistent with experimental data for various long-chain molecules (Harkins, 1952). As a gaseous monolayer is compressed on the subphase surface, the expanded (liquid-like) state will normally appear. Usually, this is accompanied by a constant pressure region in the isotherm, in which the floating film consists of a mixture of two phases. The phase change is generally thought of as a first-order thermodynamic transition, analogous to evaporation in three dimensions. Some workers have postulated the existence of monolayers in which there is a gradual transition between the gaseous and expanded states (Harkins, 1952; Gaines, 1966). This may be similar to the compression of a three-dimensional gas above its critical temperature, as shown in figure 1.6 (isotherm at temperature T_3).

2.4.2 Expanded phase

The nature of the expanded and condensed phases formed on compression of a gaseous monolayer may be more conveniently discussed by reference to the pressure versus area isotherms for specific materials. Figure 2.7 shows such a plot for n-pentadecanoic acid ($C_{14}H_{29}COOH$), measured at 25 °C and on a subphase of $10^{-2}\,\mathrm{M\,HCl}$ (Pallas and Pethica, 1985). The expanded phase (E) is evident at surface pressures below $6\,\mathrm{mN\,m^{-1}}$ (the gaseous phase was not observed in this experiment). The area per molecule in the expanded monolayer is much less than that expected for gaseous monolayers (equation

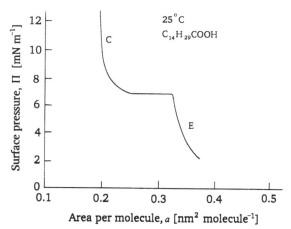

Figure 2.7 Surface pressure versus area per molecule isotherm for n-
pentadecanoic acid, at 25 °C on a subphase of 10^{-2} M HCl. (Reprinted
with permission from Pallas, N. R. and Pethica, B. A. (1985) Langmuir,
1, 509–13. Copyright 1985 American Chemical Society.)

2.5), but is still significantly greater than the area associated with the cross-
section of the cylindrically shaped fatty acid molecule ($\approx 0.19\,\text{nm}^2$). In the
expanded phase, the area per molecule varies considerably with the surface
pressure and there is no apparent relation between the observed molecular
area and the dimensions of the constituent molecules.

A plateau at a surface pressure of about $7\,\text{mN m}^{-1}$ in figure 2.7 indicates the
appearance of a condensed (C) monolayer phase. The constant pressure region
is characteristic of a first-order thermodynamic transition. There has, however,
been some debate on this point. For example, many isotherms for simple long-
chain organic compounds do not show (completely) horizontal sections for the
expanded to condensed transition. This has led to speculation that higher-order
transitions may be involved. However, it is probably more likely that constant
Π regions are not observed because of the effect of impurities (Pallas and
Pethica, 1985). The finite number of molecules involved in monolayer phase
transitions (compared to three-dimensional phase transitions) can also
considerably shorten the constant pressure plateau (Birdi, 1989).

The expanded to condensed phase changes observed in simple long-chain
compounds are affected markedly by the length of the hydrocarbon chain
and the temperature. This is illustrated by figure 2.8, which shows the
isotherms for n-pentadecanoic acid and n-hexadecanoic acid (palmitic acid –
$C_{15}H_{31}COOH$) measured at 30 °C and on a subphase of 10^{-2} M HCl (Pallas
and Pethica, 1985). Comparison with figure 2.7 reveals that an *increase* in
temperature has *increased* the surface pressure of the phase transition. A
decrease in the hydrocarbon chain length produces a similar result. Both

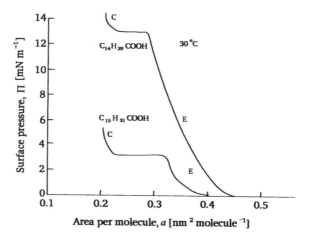

Figure 2.8 Surface pressure versus area per molecule isotherms for n-pentadecanoic acid and n-hexadecanoic acid, at 30°C on a subphase of 10^{-2} M HCl. (Reprinted with permission from Pallas, N. R. and Pethica, B. A. (1985) Langmuir, **1**, 509–13. Copyright 1985 American Chemical Society.)

effects may be understood by considering the forces between the molecules in the floating monolayer. A decrease in the length of the chain leads to decreased van der Waals' forces between the molecules, resulting in reduced cohesion within the film (with a consequent reduction in the melting point of the bulk material); in contrast, a decrease in temperature leads to less thermal motion, tending to condense the film. As a rule, reducing the hydrocarbon chain length of a long-chain fatty acid by one methylene group is roughly equivalent to a temperature increase of 5–10 K. The isotherms in figures 2.7 and 2.8 should also be contrasted with those shown in figure 1.6 for a three-dimensional substance. Further reductions in temperature (or increases in chain length) are expected to result eventually in the disappearance of the expanded to condensed phase change as the gaseous monolayer condenses directly to a solid.

There have been several attempts to develop theoretical models for two-dimensional phase transitions, such as that from an expanded to a condensed monolayer state. If it is assumed that van der Waals' forces between the hydrocarbon chains are largely responsible for the transitions in long-chain fatty acids, then the following expression may be written for the internal energy of the molecules (Jähnig, 1979; Marcelja, 1974)

$$U = E_{intra} + E_{disp} + \Pi a \qquad (2.6)$$

where E_{intra} is the internal energy of a single chain in a given conformation and E_{disp} gives the van der Waals' interaction of the chain with its neighbours

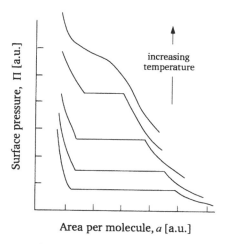

*Figure 2.9 Theoretical surface pressure versus area per molecule isotherms
for a long-chain molecule at different temperature. (After Marcelja, 1974.
Reproduced with permission from Elsevier Science.)*

(this force of attraction results from an electron correlation effect and is also
called the *London force* or *dispersion force* (Tredgold, 1994)). On this basis, a
series of theoretical surface pressure versus area isotherms may be generated
(Marcelja, 1974). As shown in figure 2.9, these curves do show qualitative
agreement with experimental data (figures 2.7 and 2.8). Modelling of
monolayers is further discussed in the book by Ulman (1991).

The coexistence of both condensed and expanded phases in floating
monolayers may be observed directly by incorporating a small fluorescent
dye probe into the film. Figure 2.10 gives the results of such an experiment
using a particular phospholipid (Lösche et al., 1988). The photographs were
taken at the points shown on the pressure versus area curve and reveal the
existence of two phases as the monolayer is compressed in the constant
pressure region of the isotherm. Nuclei of the condensed phase appear
when the plateau in the isotherm is reached (photograph a). These domains
are compact and their mean size increases with increasing surface pressure.
Finally, a hexagonal lattice is formed (photograph g).

An alternative observation method is *Brewster angle microscopy* (section
7.6). Figure 2.11 shows micrographs of the expanded to condensed phase
transitions observed by this technique for fatty acid and phospholipid
monolayers (Hönig and Möbius, 1992). While the fatty acid (photograph
(a)) shows large circular domains, the phospholipid domains (photograph
(b)) exhibit more complex shapes. Such images may be analysed to
obtain information concerning molecular orientation (Overbeck et al.,
1994a & b).

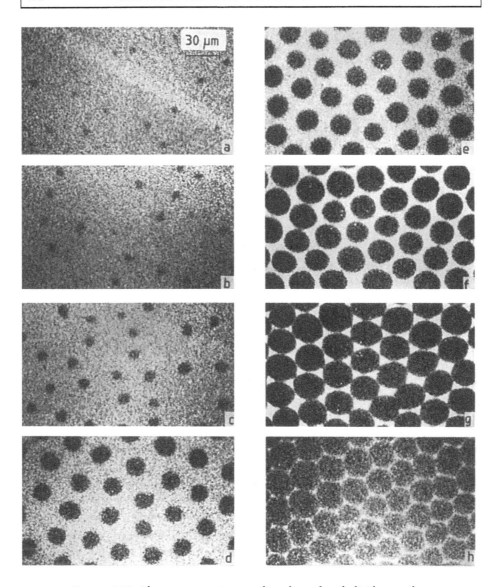

Figure 2.10 Fluorescence micrographs of a phospholipid monolayer (dimyristol-phosphatidic acid), doped with a dye, at different points on the isotherm as indicated opposite. (After Lösche et al., 1988. Reproduced with permission from Academic Press.)

2.4.3   Condensed phases

As the length of the hydrocarbon chain in a simple long-chain fatty acid is increased, the expanded state disappears and a direct transition from a gas

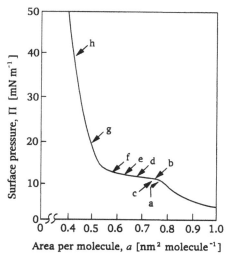

Figure 2.10 (cont.)

to a condensed phase takes place. This is illustrated by the isotherm for n-docosanoic acid (behenic acid – $C_{21}H_{43}COOH$), recorded at a temperature of 15 °C and on a subphase of 0.01 N HCl, shown in figure 2.12 (Stenhagen, 1955). Two low-pressure phases and two high pressure phases may be identified by very careful measurement.

The nomenclature used by various workers to describe these phases can be confusing (Gaines, 1966). Here, the method adopted by Harkins (1952) is used. The condensed phases include L_2, L_2' (liquid condensed), LS (super-liquid), S (solid) and CS (close-packed solid). *The use of the term 'liquid' for some of these monolayer states simply reflects the historical assumptions that the phases were liquid-like.* However, it is now known that all the condensed phases have well-defined in-plane structures and exhibit distinct X-ray diffraction peaks (Dutta, 1990).

An inspection of the abscissa in figure 2.12 reveals that the area per molecule of the various condensed monolayer phases is approaching that of the molecular cross-section. It is therefore reasonable to assume that the various monolayer states are related to different interactions, and therefore different arrangements, of the polar groups and hydrocarbon chains. Both the lower pressure L_2 and L_2' phases observed in figure 2.12 are fairly compressible. These phases have the long axes of the molecules tilted from the normal to the interface plane: in the L_2 phase, the alkyl chains are tilted towards the *nearest* neighbours and the L_2' state has chains tilted towards the *next nearest* neighbours (90° away from the L_2 tilt direction) (Shih et al., 1992a).

(a)

(b)

Figure 2.11 Brewster angle micrograph of the expanded to condensed phase transition in: (a) pentadecanoic acid (surface pressure $\approx 5\,mN\,m^{-1}$): (b) a phospholipid (dimyristoyl-phosphatidylethanolamine) ($8\,mN\,m^{-1}$). Subphase temperature $= 22\,°C$. The bars on the photographs correspond to $100\,\mu m$. (Photographs provided by Dirk Hönig and Daniela Spohn, Max-Planck-Institut für biophysikalische Chemie, Göttingen, Germany.)

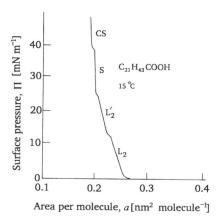

Figure 2.12 Surface pressure versus area per molecule isotherm for n-docosanoic acid, at 15 °C on a subphase of 10^{-2} N HCl. (After Stenhagen, 1955. Reproduced with permission from Academic Press.)

When a surface pressure of about 25 mN m^{-1} is reached, the floating monolayer becomes comparatively incompressible and the S phase is apparent. Further compression leads to the CS state, which eventually collapses with increasing surface pressure. The various transitions between the condensed monolayer phases are often accompanied by a (very) short region of constant pressure, possibly associated with the enthalpy change of a first order transition. For the higher pressure phases, the chain axes are normal to the interface plane.

The LS phase is characterized by a very low viscosity over a certain temperature range (Harkins, 1952). It has also been called a *rotator* phase, after the corresponding structures in lamellar paraffins (Shih et al., 1992a). This monolayer state is not observed in the isotherm in figure 2.12.

The precise nature and the molecular arrangement of the various condensed monolayer phases have been the subject of considerable speculation (Stenhagen, 1955; Gaines, 1966). It is now becoming evident that the various monolayer states are similar to mesophases found in the smectic liquid crystals described in chapter 1 (Peterson, 1992). For example, X-ray diffraction of floating fatty acid monolayers has revealed that phases L_2 and L_2' show the same unit cell symmetries as the smectic liquid crystalline phases S_I and S_F, respectively (Kenn et al. 1991). Details of these mesophases may be found in the book by Gray and Goodby (1984). Table 2.1 provides a summary of the main condensed monolayer phases that have been observed with long-chain compounds, with some of their characteristics. Transitions $L_2' \rightarrow LS$, $LS \rightarrow S$, $S \rightarrow CS$, $L_2' \rightarrow CS$ and $L_2 \rightarrow CS$ involve discontinuities in density, as expected in first-order thermodynamic phase changes.

It should be noted that monolayer states other than those listed in table 2.2 may exist. For example, smectic S_H and S_K phases have been identified

Table 2.2 *Condensed monolayer phases for long-chain organic compounds and their corresponding smectic liquid crystalline phases. (After Kenn et al., 1991.)*

Phase	Name	Smectic LC phase	Characteristics
L_2	liquid-condensed	S_I	tilted molecules
L_2'	liquid-condensed	S_F	tilted molecules, but with tilt direction at 90° to L_2 phase; similar compressibility to L_2 phase.
LS	super-liquid	hexatic S_B	low viscosity at certain temperatures; upright molecules; less compressible than L_2 and L_2' phases; similar compressibility to S and CS phases.
S	solid	S_E	upright molecules; high collapse pressure.
CS	close-packed solid	polycrystalline solid	upright molecules; high collapse pressure; lower collapse area than S phase.

for n-docosanoic acid (Bibo and Peterson, 1992). A further phase, the hexatic smectic-L appears when the n-docosanoic acid is mixed with ethyl eicosanoate. This phase is predicted to occur between the smectic S_I and S_F phases (see section 1.2).

2.4.4   Phase diagrams

If the surface pressure versus area measurements are undertaken at several temperatures, and the points corresponding to the same phase transitions are plotted on a pressure versus temperatures diagram, the resulting diagram will show the range of temperature and pressures over which the various phases exist. Such a phase diagram for n-docosanoic acid is given in figure 2.13. It is now evident why the LS phase is not observed in the isotherm shown in figure 2.12. A vertical line corresponding to 15 °C will pass through the phases L_2, L_2', S and CS. However, the LS phase will not appear in the pressure versus area diagram unless the temperature is increased to about 25 °C. It should be emphasised that the phase diagram shown in figure 2.13 is probably an oversimplification; a generalized diagram for long-chain fatty acids is given by Overbeck et al. (1994b).

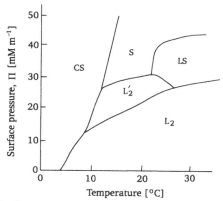

Figure 2.13 Surface pressure versus temperature phase diagram for n-docosanoic acid. (After Peterson, 1992. Reproduced with permission from Research Studies Press.)

The nature of phase diagrams for monolayer-forming materials has very important implications for the control of variables such as temperature and surface pressure in monolayer experiments (and for the transfer of monolayers to solid supports). For instance, if a monolayer is controlled so that it is in a condensed phase that is very close to a phase boundary in the phase diagram, a small change in either temperature or surface pressure may alter the state of the floating layer, affecting the deposition characteristics of the monolayer on solid supports.

The monolayer characteristics of many organic materials may be found in the comprehensive handbook edited by Mingotaud et al. (1993).

2.4.5 Surface phase rule

Gibbs' phase rule, introduced in chapter 1, may be extended to monolayer phases in equilibrium (Birdi, 1989; Gaines, 1966). Here the relationship is written

$$F = (C^b + C^s) - (P^b + P^s) + 3 \qquad (2.7)$$

where C^b is the number of components in the bulk, C^s is the number confined to the surface, P^b is the number of bulk phases and P^s is the number of surface phases in equilibrium with each other. For a monocomponent monolayer at the interface between a pure liquid and a pure gas

$$C^b = 2; \qquad C^s = 1; \qquad P^b = 2; \qquad P^s = 1.$$

There are three degrees of freedom. If the temperature and external pressure are fixed (as in most monolayer experiments), then there is only one independent variable. Consequently, the surface pressure will vary with the area of the monolayer. If there are two phases present in the monolayer (e.g., expanded and condensed), then there are only two degrees of freedom. Normally, the external pressure and temperature are already fixed. The surface pressure must therefore remain constant during the phase transformation, as shown in the isotherms in figures 2.7 and 2.8. For three surface phases to be in equilibrium simultaneously, there are no additional degrees of freedom. This is evident from the phase diagram in figure 2.13, e.g., the areas corresponding to the L_2, L_2' and LS phases meet at a point of defined temperature and surface pressure. The application of the phase rule to multicomponent systems is discussed in section 2.7.

2.5 Monolayer compressibility and viscosity

An important characteristic of a monolayer phase is its *compressibility* C. By analogy with the compressibility for a bulk material, this is defined

$$C = -\frac{1}{a}\left(\frac{\partial a}{\partial \Pi}\right)_{T,P,n_i} \tag{2.8}$$

where n is the number of moles of material. Therefore C may be evaluated directly from the slope of the surface pressure versus area isotherm. Figure 2.12 shows that the compressibility of the lower pressure L_2 and L_2' phases of n-docosanoic acid is about an order of magnitude greater than that for the more condensed S and CS phases. For the L_2 and CS phases, these values are approximately 8×10^{-3} m mN^{-1} and 3×10^{-4} m mN^{-1}, respectively (Kenn et al., 1991).

A comparison can also be made between the rheological properties of monolayer films and those of bulk matter. If the monolayer is highly rigid, it will act in a similar way to a solid when a stress is applied and deform elastically. It may be appropriate to describe this kind of behaviour using a set of elastic moduli (Gaines, 1966). In contrast, some phases of floating monolayers may be more similar to liquids and the layer will flow in response to an applied shear stress. A useful parameter to quantify this behaviour is the *coefficient of surface viscosity* η_S given by

$$(\text{shear stress}) = \eta_S \times (\text{rate of flow}) \tag{2.9}$$

The shear stress in the monolayer is simply the tangential force per unit length. The surface viscosity may be independent of shear rate (Newtonian behaviour), or may change with the rate of flow (non-Newtonian). Sometimes complex viscoelastic behaviour can result in surface pressure variations across the monolayer and consequently to compression rate dependent hysteresis in a compression–expansion isotherm measurement. In such cases it is

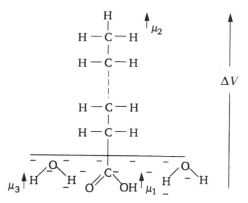

Figure 2.14 Origin of the surface potential ΔV in a long-chain fatty acid molecule at the air/water interface.

important that the nature of the compression mechanism (e.g., uniaxial, isotropic) and the compression rate are stated with the pressure versus area curve. There may also be implications for the transfer of the monolayer onto a solid support; this is discussed further in the next chapter.

2.6 Electrical effects in monolayers

Measurement of the electrical properties of the monolayer/air interface can also provide some insight into the nature of the various phases and phase changes. A convenient experimental technique is to monitor the *surface potential* ΔV. This is the potential difference produced at a point immediately above the subphase surface when a floating monolayer is introduced. It arises because monolayer-forming materials usually possess permanent electric dipoles. If the floating film is thought of as an array of dipoles whose effective dipole moments in the direction perpendicular to the subphase surface are $\mu_\perp$, the surface potential may be approximated by

$$\Delta V = \frac{n\mu_\perp}{\epsilon_0} \tag{2.10}$$

where n is the surface density of dipoles (i.e., the number of molecules per unit area) and ϵ_0 is the permittivity of free space ($\epsilon_0 = 8.854 \times 10^{-12}$ F m^{-1}). The choice of a relative permittivity of unity in equation (2.10) is arbitrary and better agreement with experimental data can be obtained by introducing an effective local permittivity (Oliveira et al., 1992). The overall dipole moment $\mu_\perp$ will include contributions from the polar head group of the molecule (μ_1), the hydrocarbon chain (μ_2) (associated with the CH$_3$ group at the end of the molecule) and the surface water molecules (μ_3), as shown in figure 2.14. For a fully condensed monolayer of n-octadecanoic acid,

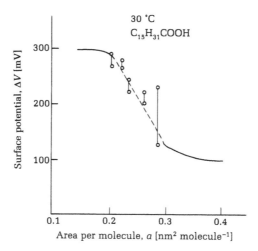

Figure 2.15 Surface potential versus area per molecule plot for n-hexadecanoic acid, at 30°C on a subphase of 10^{-2} M HCl. (Reprinted with permission from Pallas, N. R. and Pethica, B. A. (1985) Langmuir, *1, 509–13. Copyright 1985 American Chemical Society.)*

$\mu_1 = 990$ mD, $\mu_2 = 330$ mD and $\mu_3 = -65$ mD (Oliveira et al., 1992). If the monolayer becomes ionized (see below), then it is necessary to include the potential due to the double-layer (see next section).

As a monolayer of n-octadecanoic acid is compressed, the surface potential will remain close to zero until the average area per molecule is less than 0.4 nm^2 (Oliveira et al., 1992). A rapid increase in ΔV then occurs until a maximum value of several hundred millivolts is achieved. Large fluctuations in ΔV can occur for monolayers in a two phase region. These potential variations reflect the presence of patches of dense and dilute monolayer, observable by the finite size of the measuring electrode. An example is shown in figure 2.15 for n-hexadecanoic acid on 10^{-2}M HCl at 30°C (Pallas and Pethica, 1985). This curve should be contrasted with the pressure versus area curve, measured under identical conditions, shown in figure 2.8.

It is also possible to monitor the *displacement current* generated as the dipoles in the monolayer re-organize during compression (Iwamoto et al., 1992). The technique is similar to that of *thermally stimulated conductivity* (or *thermally stimulated depolarization*) of a dielectric (Harrop, 1972; Blythe, 1979). However, in the latter case, the dipole re-orientation is caused by a temperature change (see section 6.6.4).

2.6.1 Ionized monolayers

The discussion so far has been confined to monolayers of neutral molecules that bear no electrical charge. This would be the situation for a long-chain fatty acid at a pH of 4 or less. However, on making the subphase more alkaline, ionization of the polar head groups will occur to form hydrogen ions in the subphase and carboxylate ions in the film, i.e.,

$$C_nH_{2n+1}COOH \rightleftharpoons C_nH_{2n+1}CO_2^- + H^+ \qquad (2.11)$$

The pH at which half the molecules of an acid are ionized is known as the pK_A of the acid. The expected pK_A for n-octadecanoic acid on a water surface is 5.6 (Betts and Pethica, 1956). Charges associated with ionized monolayers will be neutralized by counterions in (or added to) the aqueous subphase. The distance over which this neutralization or 'screening' occurs is related to the concentration of these ions. The resulting electric potential will modify the measured surface potential. To a first approximation, this will now be given by

$$\Delta V = n\mu_\perp + \psi_0 \qquad (2.12)$$

where ψ_0 represents the maximum potential difference between the surface and the bulk of the subphase.

The exact form of the variation of the potential ψ with distance beneath the floating monolayer is complex; for example, it is likely that ψ_0 will depend on $\mu_\perp$ (Gaines, 1966). A simplified model, based on the classic Gouy–Chapman theory, is shown in figure 2.16 (Birdi, 1989). It is assumed that the monolayer is negatively charged at the air/water interface ($x = 0$ in the figure). Positively and negatively charged ions in the subphase will then be distributed to achieve charge neutrality. This distribution will be governed by *Boltzmann statistics*, resulting in a variation of ψ from a maximum value of ψ_0 at the surface to zero in the bulk of the subphase. The distance over which ψ changes is known as the *diffuse* or *Gouy–Chapman double layer* ($x = 0$ to $x = L$ in figure 2.16). In the real situation specific ions may be adsorbed on the polar ends of the mono-layer or may even penetrate this layer; furthermore ions in solution are likely to be *solvated* (surrounded by oriented water molecules) and can approach the surface only to a distance about equal to the radius of the solvated ions.

One important consequence of the above is that the pH of the surface region will differ from that of the bulk (Gaines, 1966)

$$pH_s = pH_b + \left(\frac{q\psi}{2.3kT} \right) \qquad (2.13)$$

where pH_s is the surface pH, pH_b is that of the bulk solution, and q is the electronic charge (1.60×10^{-19} C).

Langmuir–Blodgett films of long-chain fatty acids are often prepared by *deliberately* adding divalent ions to the subphase to improve the deposition characteristics of the monolayer film. The floating layer will be a mixture of the fatty acid and the fatty acid salt. The salt concentration in the

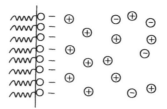

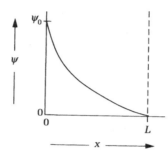

Figure 2.16 Top: simple model for the charge distribution at an air/mono-layer/water interface. Bottom: variation of potential ψ with distance.

monolayer will depend upon the pH. In a subphase containing cadmium chloride, the following reaction will take place

$$2C_nH_{2n+1}COOH + CdCl_2 \rightleftharpoons (C_nH_{2n+1}CO_2)_2^- Cd^{2+} + 2HCl \qquad (2.14)$$

The salt formation is favoured by high subphase pH. In the specific case of a floating monolayer of n-eicosanoic acid (arachidic acid) on a subphase at room temperature containing a cadmium salt in a concentration of 10^{-4} M and having a pH = 5.7, a monolayer comprising about 50% cadmium eicosenoate and 50% eicosanoic acid will be formed (Blodgett and Langmuir, 1937).

Figure 2.17 shows the effect of the incorporation of calcium ions on the pressure versus area isotherms for n-heneicosanoic acid (Shih et al., 1992b). The isotherms have been displaced horizontally for clarity. At the lowest pH (2.1), the isotherm is identical to that observed at the same temperature without ions in the subphase. The L_2 and L_2' phases, together with the transition to a condensed monolayer state, are evident. However, as the pH is increased, the structure in the isotherms corresponding to the phase transitions is lost. At the highest pH (10.4), the surface pressure versus area curve for the salt monolayer is featureless. The general forms of the isotherms in figure 2.17 are observed for many long-chain fatty acid/salt combinations. The formation of the salt monolayer appears to convert the floating mono-layer into a more condensed solid as the head groups are drawn closer together and the tails move closer to the vertical (Shih et al., 1992b). Theoretical studies

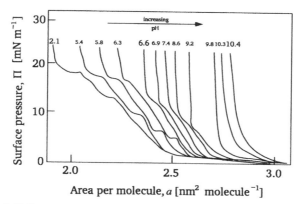

Figure 2.17 Pressure versus area per molecule isotherms, at different pH values, for monolayers of n-heneicosanoic acid spread on a solution of 0.5 mN calcium chloride. The isotherms have been displaced horizontally for clarity; the abscissa is valid only for the isotherm at pH 2.1. (After Shih et al., 1992b. Reproduced with permission from the American Institute of Physics.)

also show that the role of ions in fatty acid monolayers is to provide long-range attractive forces that cause a two-dimensional lattice to form, even at low amphiphile concentrations (Modak and Datta, 1994).

Addition of ions to the subphase can also affect the mechanical properties of the surface monolayer. Traces of triply charged cations in particular (e.g., Fe^{3+}, Al^{3+}) can give rise to extremely rigid layers (Hann, 1990).

2.7 Mixed monolayers

Interest in molecular electronics and molecular engineering has led to many studies of mixed monolayer systems. Sometimes a fatty acid is simply used to 'stabilize' a compound that would not form a monolayer by itself. The resulting layers could be intimate mixtures, as shown in figure 2.18(a), or a multiphase mixture (figure 2.18(b)). The precise arrangement may usually be identified by studying an appropriate monolayer property (e.g., collapse pressure or compressibility) over a wide composition range. If the individual components are immiscible, then the film may be thought of as consisting of two (or more) separate monolayers. The area occupied by the film will be the sum of the areas of the separate films, i.e., for a two-component mixture

$$a_{12} = x_1 a_1 + x_2 a_2 \qquad (2.15)$$

where a_{12} is the average molecular area in the two-component film, x_1 and x_2 are the mole fractions of the components, and a_1 and a_2 are the molecular

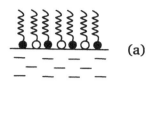

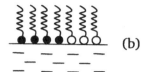

Figure 2.18 Different arrangements for the molecules in a two-component monolayer: (a) complete miscibility; (b) immiscible materials.

areas in the two single component films at the same surface pressure. For an 'ideal' mixed film, a plot of a_{12} versus x_1 should be a straight line. Any deviation from this relationship provides evidence for miscibility. Similar scaling rules may be used for other properties such as surface potential and compressibility.

Using the surface phase rule (equation (2.7)) for two components in the monolayer

$$F = 5 - P^s \tag{2.16}$$

If there is only one surface phase (i.e., the two components are completely miscible), there are four degrees of freedom: temperature, the external pressure, the monolayer surface pressure and the composition of the film. However, if the monolayer components do not mix, two surface phases are formed and one degree of freedom is removed.

If the mixed monolayer is compressed until it collapses, an additional bulk phase will be formed as monolayer material is squeezed out of the floating layer (i.e., $P^b = 3$). Now

$$F = 4 - P^s \tag{2.17}$$

For one surface phase, there are three degrees of freedom, and if external pressure and temperature are constant, the surface pressure must vary with the composition. This surface pressure will be the equilibrium spreading pressure of the mixed film (i.e., the pressure at which the monolayer is in equilibrium with the bulk phase). In the case where the monolayer components are immiscible there are two surface phases ($P^s = 2$) and only two degrees of freedom. At a fixed temperature and external pressure, the equilibrium spreading pressure will be constant, independent of composition.

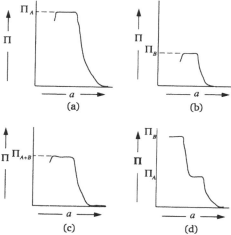

Figure 2.19 Surface pressure versus area per molecule isotherms for: (a) compound A; (b) compound B; (c) mixed monolayer of A and B with complete miscibility; (d) mixed monolayer of A and B with no miscibility.

The component of the mixed film that has the lower equilibrium spreading pressure will be squeezed out of the monolayer at the same surface pressure, whatever the initial composition of the monolayer. A measurement of equilibrium spreading pressure as a function of monolayer composition should therefore be a good test for monolayer homogeneity (see figure 2.19). Very often the collapse pressure is taken as a guide to the equilibrium spreading pressure. However, the results will require careful interpretation as the monolayer film may not be in a state of true thermodynamic equilibrium when collapse occurs.

2.8 Chemistry in monolayers

Several kinds of chemical interaction can take place between the monolayer film and one (or both) of the adjacent bulk phases. These include the oxidation of $C=C$ bonds, hydrolysis, photochemical reactions (e.g., *cis–trans* isomerism) and polymerization. These subjects have been discussed by Bubeck (1988) and Gaines (1966).

References

Adamson, A. W. (1982) *Physical Chemistry of Surfaces*, 4th edition, John Wiley, New York

Betts, J. J. and Pethica, B. A. (1956) The ionization characteristics of monolayers of weak acids and bases, *Trans. Farad. Soc.*, **52**, 1581–1589

Bibo, A. M. and Peterson, I. R. (1992) Defect annealing in monolayers displaying a smectic-L phase, *Thin Solid Films*, **210/211**, 515–18

Birdi, K. S. (1989) *Lipid and Biopolymer Monolayers at Liquid Interfaces*, Plenum Press, New York

Blodgett, K. B. and Langmuir, I. (1937) Built-up films of barium stearate and their optical properties, *Phys. Rev.*, **51**, 964–982

Blythe, A. R. (1979) *Electrical Properties of Polymers*, Cambridge University Press, Cambridge

Bubeck, C. (1988) Reactions in monolayers and Langmuir–Blodgett films, *Thin Solid Films*, **160**, 1–14

Dutta, P. (1990) Phase transitions in lipid monolayers on water: new light on an old problem, in *Phase Transitions in Surface Films 2*, eds. H. Taub, G. Torzo, H. J. Lauter and S. C. Fain, Jr., NATO ASI Series, Physics Vol. 267, Plenum Press, New York

Gaines, G. L., Jr. (1966) *Insoluble Monolayers at Liquid–Gas Interfaces*, Wiley-Interscience, New York

Gray, G. W. and Goodby, J. W. G. (1984) *Smectic Liquid Crystals*, Leonard Hill, Glasgow

Hann, R. A. (1990) Molecular structure and monolayer properties, in *Langmuir–Blodgett Films*, ed. G. G. Roberts, Plenum Press, New York

Harkins, W. D. (1952) *Physical Chemistry of Surface Films*, Reinhold, New York

Harrop, P. J. (1972) *Dielectrics*, Butterworths, London

Hönig, D. and Möbius, D. (1992) Reflectometry at the Brewster angle and Brewster angle microscopy at the air–water interface, *Thin Solid Films*, **210/211**, 64–8

Iwahashi, M., Maehara, N., Kaneko, Y., Seimiya, T., Middleton, S. R., Pallas, N. R. and Pethica, B. A. (1985) Spreading pressures for fatty-acid crystals at the air/water interface, *J. Chem. Soc., Faraday Trans.*, **81**, 973–81

Iwamoto, M., Majima, Y. and Watanabe, A. (1992) Detection of the reorganization of monolayers at the air-water interface by displacement current measurement, *Thin Solid Films*, **210/211**, 101–4

Jähnig, F. (1979) Molecular theory of lipid membrane order, *J. Chem. Phys.*, **70**, 3279–90

Kenn, R. M., Böhm, C., Bibo, A. M., Peterson, I. R., Möhwald, H., Als-Nielsen, J. and Kjaer, K. (1991) Mesophases and crystalline phases in fatty acid monolayers, *J. Phys. Chem.*, **95**, 2092–2097

Lösche, M., Duwe, H. P. and Möhwald, H. (1988) Quantitative analysis of surface textures in phospholipid monolayer phase transitions, *J. Colloid Interface Sci.*, **126**, 432–44

Marcelja, S. (1974) Chain ordering in liquid crystals II. Structure of bilayer membranes, *Biochem. Biophys. Acta*, **367**, 165–76

Mingotaud, A-F., Mingotaud, C. and Patterson, L. K., eds. (1993) *Handbook of Monolayers*, Academic Press, Orlando

Modak, S. and Datta, A. (1994) Role of ionic head groups in the ''zero-pressure '' phase of fatty acid monolayers, *J. Chem. Phys.*, **98**, 1–3

Oliveira, O. N., Jr., Taylor, D. M. and Morgan, H. (1992) Modelling of surface potential-area dependence of a stearic acid monolayer, *Thin Solid Films*, **210/211**, 76–8

Overbeck, G. A., Hönig, D., Wolthaus, L., Gnade, M. and Möbius, D. (1994a) Observation of bond orientational order in floating and transferred monolayers with Brewster angle microscopy, *Thin Solid Films*, **242**, 26–32

Overbeck, G. A., Hönig, D. and Möbius, D. (1994b) Stars, stripes and shells in monolayers: simulation of the molecular arrangement in Schlieren structures, *Thin Solid Films*, **242**, 213–19

Pallas, N. R. and Pethica, B.A. (1985) Liquid-expanded to liquid condensed transitions in lipid monolayers at the air/water interface, *Langmuir*, **1**, 509–13

Peterson, I. R. (1992) Langmuir–Blodgett Films, in *Molecular Electronics*, ed. G. J. Ashwell, pp. 117–206, Research Studies Press, Taunton

Shih, M. C., Bohanon, T. M., Mikrut, J. M., Zschack, P. and Dutta, P. (1992a) X-ray diffraction study of the 'superliquid' region of the phase diagram of a Langmuir monolayer, *Phys. Rev. A.*, **45**, 5734–7

Shih, M. C., Bohanon, T. M., Mikrut, J. M., Zschack, P and Dutta, P. (1992b) Pressure and pH dependence of the structure of a fatty acid monolayer with calcium ions in the subphase, *J. Chem. Phys.*, **96**, 1556–9

Stenhagen, E. (1955) Surface films, in *Determination of Organic Structures by Physical Methods*, eds. E. A. Braunde and F. C. Nachod, pp. 325–71, Academic Press, New York

Tredgold, R. H. (1994) *Order in Thin Organic Films*, Cambridge University Press, Cambridge

Ulman, A. (1991) *Ultrathin Organic Films*, Academic Press, San Diego

Walton, A. J. (1976) *Three Phases of Matter*, McGraw-Hill, Maidenhead

3

Film deposition

3.1 Deposition principles

The Langmuir–Blodgett (LB) technique, first introduced by Irving Langmuir and applied extensively by Katharine Blodgett, involves the vertical movement of a solid *substrate* through the monolayer/air interface (Blodgett, 1934; Blodgett, 1935; Langmuir, 1920). Blodgett's and Langmuir's original papers contain a wealth of useful experimental advice and are still excellent starting points for anyone considering working in the area today (Blodgett, 1935; Blodgett and Langmuir, 1937).

The surface pressure and temperature of the monolayer are first controlled so that the organic film is in a condensed and stable state. For fatty acid type materials, deposition generally proceeds from either the L_2', LS or S phase (with surface pressures in the range 20–40 mN m^{-1} and temperatures 15–20 °C). However, it is also possible to start from one of the other monolayer states. The molecular organization in the resulting LB film will depend on these initial conditions.

Figure 3.1 shows the commonest form of LB film deposition. The substrate is hydrophilic and the first monolayer is transferred, like a carpet, as the substrate is raised through the water. The substrate may therefore be placed in the subphase before the monolayer is spread. Subsequently, a monolayer is deposited on each traversal of the monolayer/air interface. As shown, these stack in a head-to-head and tail-to-tail pattern; this deposition mode is called *Y-type*. Although this is the most frequently encountered situation, instances in which the floating monolayer is only transferred to the substrate as it is being inserted into the subphase, or only as it is being removed, are often observed. These deposition modes are called *X-type* (monolayer transfer on the downstroke only) and *Z-type* (transfer on the upstroke only) and are illustrated in figure 3.2. Mixed deposition modes are sometimes encountered

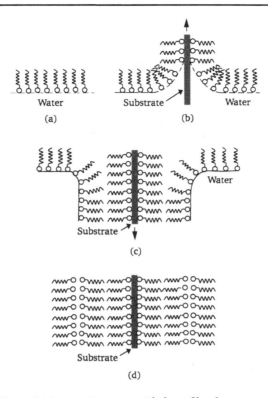

Figure 3.1 Y-type Langmuir–Blodgett film deposition.

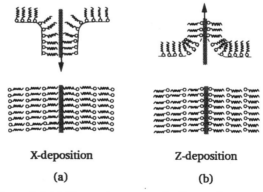

Figure 3.2 (a) X-type deposition; (b) Z-type deposition.

and, for some materials, the deposition type can change as the LB film is built up. It is also possible to build up films containing more than one type of monomolecular layer. In the simplest case, alternate-layer films may be produced by raising the substrate through a monolayer of one material (consisting of molecules of compound A, say) and then lowering the substrate

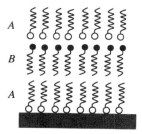

Figure 3.3 An alternate-layer Langmuir–Blodgett film built up from monolayers of compound A and monolayers of compound B.

through a monolayer of a second substance (compound *B*). A multilayer structure consisting of *ABABAB . . .* layers is produced, figure 3.3.

Film transfer is characterized by measurement of the *deposition ratio*, τ (also called the *transfer ratio*). This is the decrease in the area occupied by the monolayer (held at constant pressure) on the water surface divided by the coated area of the solid substrate, i.e.,

$$\tau = \frac{A_\text{L}}{A_\text{S}} \tag{3.1}$$

where A_L is the decrease in the area occupied by the monolayer on the water surface and A_S is the coated area of the solid substrate. If asymmetric substrates are used (e.g., a glass slide metallized on one surface) then it is unlikely that τ will be identical for both surfaces. Transfer ratios significantly outside the range 0.95 to 1.05 suggest poor film homogeneity. The accurate measurement of τ for alternate-layer films can present a problem as, in many alternate-layer troughs, the substrate holder is passed through the floating monolayer (section 3.7.1). Fatty acid and fatty acid salt monolayers normally deposit as Y-type films. However, X-type deposition is possible with suitable changes in the dipping conditions: X-type deposition is favoured by high pH values. Appropriate choice of dipping conditions also enables long-chain esters to be deposited as either X- or Y-type layers (Petty and Barlow, 1990). There are many reports of Z-type films. Typically, these concern aromatic molecules with short or no hydrocarbon chains that do not form true *monolayers* at the air/water interface. Such materials are very different from the classic fatty acid type compounds discussed in the previous chapter.

The diagrams shown in figures 3.1 to 3.3 are simple sketches. These may not accurately represent the real arrangements of molecules on a solid surface. For LB films of *fatty acid salts*, X-ray diffraction experiments show that the long axes of the molecules in the LB film are indeed orthogonal to the substrate plane, as shown in most 'molecular stick' drawings (Petty, 1990).

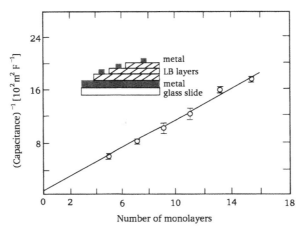

Figure 3.4 Reciprocal capacitance per unit area versus number of monolayers for Au/cadmium stearate/Al structures. The inset shows the
structure of the metal/LB film/metal capacitors. (After Batey, 1983.)

The molecular arrangement in the deposited film is therefore similar to that
of the LS, S and CS condensed monolayer phases on the water surface.
However, the long molecular axes in many simple *fatty acid* LB layers are
inclined at an angle to the substrate normal, the tilt angle depending on the
precise deposition conditions. For example, the molecular tilt elevation of
fatty acid films deposited from the L_2' phase can vary with the deposition
pressure (Peterson, 1992). This tilt elevation may also change from layer to
layer.

The reproducibility of the LB deposition process is monitored by measuring
a suitable physical characteristic of the organic film as a function of time.
Measurements of the film thickness, its optical density or the frequency
shift of a quartz crystal oscillator are all straightforward. Perhaps the most
stringent test is to sandwich the multilayer film between two metal electrodes
(chapter 6) and to monitor the capacitance of this structure as a function of
the number of dipping cycles (Peterson, 1990 & 1992). Since capacitance
varies as the reciprocal of thickness, a plot of reciprocal capacitance of
the metal/LB film/metal structure versus the number of dipping cycles
should be a straight line. Figure 3.4 shows such a plot for cadmium stearate
layers sandwiched between evaporated aluminium and gold electrodes
(Batey, 1983). The linear form of the graph confirms the reproducibility of
monolayer capacitance and therefore of the film deposition from one monolayer to the next. The slope of the straight line is related to the permittivity
and thickness of each monolayer, while the intercept on the ordinate yields
similar information about the oxide layers on the metal electrodes (Petty,
1990).

3.2 Deposition speed

As the substrate is lowered into the subphase, it can be moved quite rapidly without affecting the monolayer transfer. However, on withdrawal through the floating monolayer it is important not to raise the substrate faster than the rate at which water drains from the solid. This drainage is not due to gravity but is a result of the adhesion between the monolayer being transferred and the material on the substrate which acts along the line of contact and so drives out the water film. The rate at which films can be built-up is limited by the rate at which the ascending substrate sheds water. It is normal to transfer the initial monolayer onto a solid substrate relatively slowly: speeds of $10\,\mu\mathrm{m\,s^{-1}}$ to a few $\mathrm{mm\,s^{-1}}$ are typical. However, much faster speeds, up to several $\mathrm{cm\,s^{-1}}$, are possible once the initial layer has been transferred. To improve film deposition, it is often appropriate to temporarily halt the dipping process after an upstroke and wait until the transferred monolayer is completely dry before continuing the deposition cycle.

3.3 Molecular re-arrangement

The final molecular arrangements in an LB layer may not be always as shown in figures 3.1 to 3.3. For fatty acids, early experiments using X-ray diffraction (chapter 5) revealed that *the spacing between the hydrophilic head groups was nearly the same, and equal to twice the length of the hydrocarbon chain, whether they were deposited as X-type or Y-type films*. It is also becoming clear from other work (e.g., neutron reflection from deuterated multilayers – Grundy et al., 1990) that some molecular layers in LB films re-arrange, during or shortly after deposition.

The driving force for the re-arrangement of molecules during LB deposition must originate from interactions such as those between the polar heads, between the hydrocarbon tails, between the head groups of one layer and the hydrocarbon tails of the other or between the heads and tails of the molecules and the water subphase. Figure 3.5 shows a possible model that accounts for most of the experimental observations in fatty acid films (Honig, 1973). On the downward movement of the substrate in X-deposition, some transferred molecules re-orient and arrange themselves so that their polar heads are next to the polar heads of the deposited layer (i.e., forming a bilayer). If half the molecules do this, then the outer surface of the multi-layer system will be hydrophobic under water and no film will transfer on the substrate upstroke. Overturning of less than half the molecules will result in partial pick-up on the upstroke (i.e., $\tau < 1$), as noted for many LB materials. The resulting multilayer film will be quite irregular in thickness, with the spacing between head groups characteristic of Y-type layers.

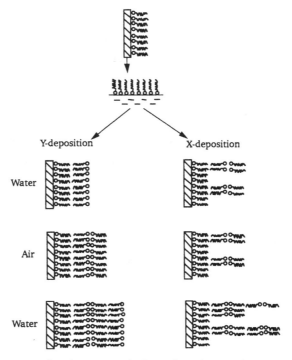

Figure 3.5 Y- and X-deposition of a long-chain fatty acid showing how re-arrangement of the X-type film can produce a similar layer spacing to that of the Y-film. (After Honig, 1973. Reproduced with permission from Academic Press.)

The microscopic processes occurring during molecular re-arrangement are unclear. One question that needs to be answered is why do long-chain fatty acid compounds transfer as Y-type layers at neutral pHs, but as X-type films if the subphase is made alkaline? The monolayer composition (salt : acid ratio) and the various interaction energies outlined above will change with the subphase pH as the charged state of the monolayer is altered. The equilibrium interaction between the ionized fatty acid and ions in the subphase is certainly more complex than that given by equation (2.14). Under normal LB deposition conditions, it is expected that the subphase will also contain anions (e.g., OH^-) which may become incorporated into the multilayer (Stephens, 1972), e.g.,

$$C_nH_{2n+1}COOH + M^{2+} + A^- \rightleftharpoons C_nH_{2n+1}COO-M-A + H^+ \qquad (3.2)$$

where M^{2+} and A^- represent the subphase cations and anions, respectively. The overturning or re-arrangement of LB film layers is likely to be influenced by such effects.

3.4 First layer effects and epitaxy

During LB film deposition, the first monomolecular layer will be transferred onto a solid substrate of a different material. This is an example of *heterogeneous crystal growth* (Neugebauer, 1970). For subsequent monolayers, transferring onto existing film, the deposition will be *homogeneous*. It is probable that the chemical and physical structure of the first monolayer will be different to that of subsequent layers. There is now much evidence to support this. For example, infrared investigations of fatty acid films (section 5.5) show that the first monolayer possesses a hexagonal packing of the CH_2 subcells in the alkyl chain, with the chain axis oriented perpendicular to the substrate surface (Bonnerot et al., 1985; Kimura et al., 1986). As the film thickness increases, a transition occurs to a structure in which the subcells have an orthorhombic packing and the main cells are packed in a monoclinic crystallographic form, with the axes of the alkyl chains inclined at an angle of 20° to 30° to the substrate normal. Reflection electron diffraction experiments (section 5.4.3) also show a tilt elevation for the long axis of fatty acid molecules which varies with the thickness of the LB film (Jones et al., 1986).

For fatty acid salt materials deposited onto metallic substrates, there may be an ion exchange between the fatty acid salt and the thin oxide layer on the substrate surface (Petty, 1990); e.g., when a calcium stearate film is transferred onto an aluminium plate, a layer of aluminium stearate is formed. Consequently, a strong chemical bond can anchor the polar head of the first LB layer to the substrate surface. Anyone who has experience of LB film deposition will be aware of this. The outer layers in a multilayer film of a long-chain fatty acid type material can be removed from the supporting surface by simple mechanical means (e.g., wiping with a tissue). This is because the film is held together by relatively weak van der Waals' forces. However, the first layer deposited remains (and can be revealed by breathing onto the apparently bare substrate).

In all forms of thin film growth, it is important to know whether the deposition is *epitaxial*, i.e., is the orientation of the molecules in the thin film the same as that of the molecules in the underlying substrate? For fatty acids, evidence from electron diffraction shows that each monolayer has the same local orientation of its molecular lattice as that of the underlying monolayer; however, this does not necessarily mean that translational order also extends from layer to layer (Peterson, 1990).

LB films of fatty acids consist of 'domains' with in-plane dimensions ranging from several hundred micrometres to a few tenths of a millimetre. If the long axes of the molecules are tilted with respect to the substrate normal, then the films will be birefringent (appendix B) and the domain structure can be seen by observation under a polarizing microscope. Figure 3.6 shows such

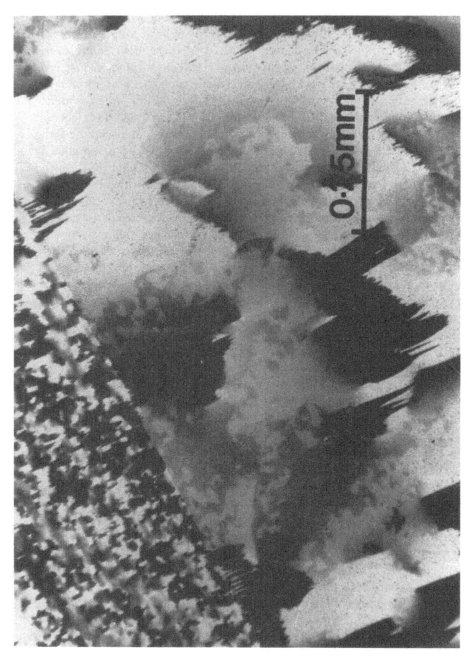

Figure 3.6 Photomicrograph of 22-tricosenoic acid (170 layers) deposited onto single crystal silicon. The substrate, apart from that shown in the top left-hand corner, was initially covered with a bilayer of n-octadecanoic acid. (After Peterson, I. R. and Russell, G. J. (1985) British Polymer Journal, 17, 364–7. Reproduced with permission from the Society of Chemical Industry.)

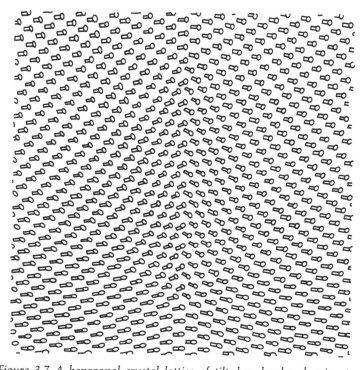

Figure 3.7 A hexagonal crystal lattice of tilted molecules showing two disclinations of opposite signs joined by a twin boundary. (After Peterson, 1990. Reproduced with permission from the Institute of Physics.)

an image for a 170-layer LB film of 22-tricosenoic acid (chapter 4) deposited onto an etched silicon substrate (Peterson and Russell, 1985). Most of the area in the photograph was initially coated with a bilayer of n-octadecanoic (stearic) acid. However, the region in the top left corner of the photograph was not coated, providing clear evidence that the 22-tricosenoic acid had developed the domain features of the underlying n-octadecanoic acid bilayer.

3.5 Defects

If the features in figure 3.6 that look like grain boundaries are studied closely, it becomes apparent that these do not form closed loops, as expected from polycrystalline layers. The defects are like those found in smectic liquid crystals (the similarity between the states of floating monolayers and smectic liquid crystals has already been noted in chapter 2). These are called *disclinations* and are points where there is a discontinuity in the orientational ordering of the molecules (Tredgold, 1994). An example is shown in figure 3.7, which shows a locally hexagonal closed packed lattice of tilted molecules

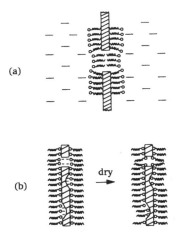

Figure 3.8 Deposition of LB layers onto rough and porous surfaces: (a) formation of a black lipid bilayer under water; (b) deposited LB monolayer in air before and after drying.

with two disclinations of opposite signs joined by a twin boundary (Peterson, 1990).

Disclination densities of about $10 \, \text{mm}^{-2}$ are observed in fatty acid LB layers. This may be reduced by annealing the monolayer on the surface before deposition or cycling between two values of surface pressure, so that the monolayer is taken between two different states (Bibo and Peterson, 1992). By processing the initial monolayer for a few hours, a high yield of $1 \, \text{mm}^2$ metal/LB film/metal cells free from disclinations may be achieved.

Other types of defect also occur in LB films. These include pin-holes and regions containing collapsed monolayer (normally observed as striations in the LB film). Careful preparation is necessary to reduce these to a minimum (section 3.8).

3.6 Deposition onto rough or porous surfaces

The surfaces of actual substrates that are used for LB deposition are likely to be rough on a molecular scale; they may even contain large pits or holes. The monolayer may be deposited conformally, preserving the surface variations of the underlying substrate, or it may span the voids and other defects in the substrate. *Under water* lipid layers can bridge small holes; a bilayer formed in this way is called a *black lipid membrane* and is used for ion transport studies (figure 3.8(a)). What happens when LB layers are deposited onto solid substrates and stored in air is less clear. Experiments show that LB multilayer films (particularly polymer materials) can be used to reduce the permeation of gases through porous membranes (Bruinsma et al., 1992).

However, this may be achieved by the organic layer collapsing into the pores. If the substrate is very irregular, then it is possible that the LB film can follow the undulations (the situation would be similar to that of laying a carpet over a mountain range). For surface roughness on the nanometre scale, it is more likely that, at the moment of deposition, the monolayer will bridge over voids, supported by a layer of water. When this layer has drained or dried, the film will collapse, as shown in figure 3.8(b) (Gaines, 1966).

Another important feature of LB layers is apparent from figure 3.8. Under the water the polar ends of the molecules will normally be outermost; however, in air the alkyl chains will be on the outside, making the film hydrophobic (section 3.9.2).

3.7 Equipment for LB deposition

3.7.1 Trough and compression system

Many different approaches have been made to the design of equipment for the deposition of LB films. The first requirement is for a subphase container – called a Langmuir or a Langmuir–Blodgett *trough*. The material used for all parts of the trough that come into direct contact with the subphase should be inert and able to withstand the organic solvents used for monolayer spreading and cleaning. A popular choice is PTFE; co-polymers of this material are generally more easily formed and are also used.

In the simplest type of equipment, the trough forms an integral part of the monolayer barrier. The principle of operation is shown in figure 3.9(a). The barrier may be moved via a suitable gearing system to an electric motor. Most of the trough need not be very deep (a few millimetres) as a 'well' can allow for the transfer of the floating layer onto a substrate (figure 3.9(b)). Most LB troughs are based on rectangular geometries, but systems designed in a circular fashion are also used.

A continuous flexible belt system (usually made from PTFE-coated glass fibre) may also be used to contain and compress the floating monolayer (Petty and Barlow, 1990). This avoids leakage problems, also the need for careful water-level adjustment. As this barrier is independent of the trough containing the subphase, it is readily removed for cleaning. Figure 3.10 shows two configurations for such systems (Malcolm, 1988). In figure 3.10(a), the barrier is stretched around six PTFE rollers. These are secured to two mobile overarms which move symmetrically inwards or outwards. The narrow channels, formed when the monolayer area is reduced, are pinched off by the roller/barrier geometry (certain monolayer materials do not flow easily into these channels). The alternative compression system shown in figure 3.10(b) facilitates the flow of monolayer material from the subphase surface onto the solid substrate (section 3.7.2).

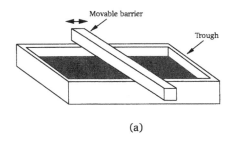

(a)

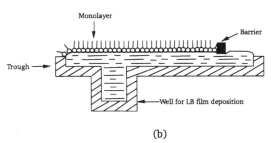

(b)

Figure 3.9 (a) Simple LB trough with a single movable barrier. (b) Cross-section of a trough with a well to facilitate LB film deposition.

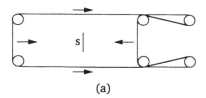

(a)

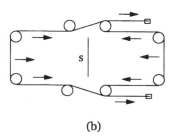

(b)

Figure 3.10 Alternative configurations for continuous perimeter belt system used to compress monolayer: (a) normal configuration; (b) arrangement with symmetrical compression and monolayer withdrawn from full width of the trough. The size and position of the substrate S is indicated. (After Malcolm, 1988. Reproduced with permission from the Institute of Physics.)

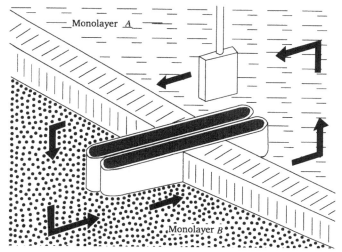

Figure 3.11 Schematic diagram for alternate-layer film deposition system.

In a further design, referred to as a 'moving-wall' trough, the side wall barriers move at the same speed as the monolayer during deposition (Kumehara et al., 1989). Furthermore, the trough width is designed to be the same as that of the solid substrate, preventing shear flow of the monolayer. This type of LB trough may have advantages for the transfer of certain monolayer films (Suga et al., 1994).

Figure 3.11 shows a schematic diagram of a system that can be used for alternate-layer deposition. A fixed barrier separates two compartments containing two different monolayer materials, compound A and compound B, floating on the water surface. A gate arrangement, built into this barrier, permits the substrate holder to pass through. Alternate-layer multilayer films of the composition $ABABAB$ . . . (i.e., as illustrated in figure 3.3) may be built up by passing the substrate up through a condensed monolayer of A, through the air, down through a condensed monolayer of B and then underwater (with the substrate holder passing through the gate) and up through layer A again. A double gate arrangement can reduce mixing of the two monolayer areas: contamination collects in the uncompressed region between the gates and is periodically removed when this surface is cleaned. Alternate-layer troughs based on a 'diamond-barrier' design have also been developed, with good monolayer flow patterns (see section 3.7.2) (Miller and Stone, 1992; Miller and Rhoden, 1994).

Many variations on simple troughs and compression systems have been proposed and some of these are used by commercial manufacturers of LB equipment (Petty and Barlow, 1990). A system for the continuous deposition of monolayer films has been demonstrated (Barraud and Vandevyver, 1983).

Here the traditional order of LB deposition, usually distributed in time (i.e., monolayer spreading, monolayer compression and film deposition) is distributed in space, all of the operations taking place simultaneously in different compartments of the trough. An automated system using a flowing subphase as a means of compression may also be useful for large-scale industrial LB film deposition (Lu et al., 1994).

3.7.2 Monolayer flow

As the floating monolayer is removed from the water surface, other material must flow in to replace it on the surface. The motion of the floating monolayer during the process of deposition will depend on both its viscosity and the barrier geometry. A highly viscous monolayer will be unable to adjust itself to maintain a homogeneous film close to a rapidly moving substrate. Standard designs of Langmuir troughs can subject molecular monolayers to very nonuniform shear forces (Malcolm, 1988, 1989a & b; Byatt-Smith and Malcolm, 1994). These arise from viscous drag originating at the sides of the trough and from the inward flow of monolayer caused by the advancing barriers. In a continuous flexible belt system, the movement of the belt draws monolayer along with it and can cause distortions in both the floating monolayer and the transferred film. The flow patterns of the surface monolayer can be observed by sprinkling fine particles directly onto the surface (Hann, 1990; Malcolm, 1989a & b). Figure 3.12 shows the result of an experiment to compare the deposition patterns for different compression geometries (Malcolm, 1988). Lines of sulphur, applied to the surface of the floating condensed monolayer (barium stearate), were transferred onto the substrate (a piece of filter paper) as it was drawn through the monolayer/air interface. The figure shows the patterns produced using (a) symmetrical compression with two moving barriers, (b) compression from one side of the substrate only (the side facing the moving barrier) and (c) using the constant perimeter barrier pattern shown in figure 3.10(b).

A surprising result is that symmetrical compression of a floating monolayer does not necessarily reduce the extent of deformation of the deposited LB layer (figure 3.12(a)). This is because asymmetric compression leads to monolayer flow around the substrate and a reduction of convergent flow on the substrate surface facing the moving barrier. The use of the continuous belt system and the withdrawal of the substrate from the full width of the trough also results in a deposited monolayer almost free from distortion (figure 3.12(c)).

3.7.3 Water quality

The water subphase itself must be of the highest purity available. In most monolayer experiments the quantity of monolayer-forming material that is

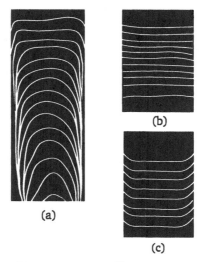

Figure 3.12 Deposition patterns on filter paper substrates for barium stearate monolayers: (a) trough with two moving barriers, symmetrical compression; (b) as (a) but compression by one barrier only (side facing moving barrier); (c) symmetrical compression using the system shown in figure 3.10(b). (After Malcolm, 1988. Reproduced with permission from the Institute of Physics.)

used is in the microgram range; the subphase volume can be 100s of milli-litres. Therefore, p.p.b. (parts per 10^9) impurities, if they are surface active, can cause a problem. Both distilled and deionized water have been used by different research groups. There is no consensus about which is better. A deionizing water system, using mixed anion and cation ion-exchange resins, with a reverse osmosis unit and ultraviolet-sterilizer (as used in the semiconductor industry) is the choice of many LB workers. Such systems can produce water of high specific resistance ($1.8 \times 10^5 \, \Omega \, \mathrm{m}$) and total organic contamination (TOC) in the p.p.b. level. Of course, once it has been produced, the handling of the water is also important. On exposure to the atmosphere the water will absorb carbon dioxide and its pH will decrease over a period of a few hours.

3.7.4 Surface pressure measurement

The two most common methods for monitoring the surface pressure are the Langmuir balance and the Wilhelmy plate. Both have similar sensitivities ($\approx 10^{-3} \, \mathrm{mN \, m^{-1}}$), but the use of the Wilhelmy plate technique is perhaps the more popular. An absolute measurement of Π is made by suspending a plate from a sensitive balance in the monolayer. Figure 3.13 shows the experimental arrangement. The forces acting on the plate are due to gravity

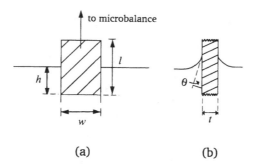

Figure 3.13 A Wilhelmy plate: (a) front view; (b) side view.

and surface tension downwards and buoyancy, due to displaced water, upwards. For a rectangular plate of dimensions l, w and t and of material of density ρ_W immersed to a depth h in a liquid of density ρ_L, the net downward force F is given by

$$F = \rho_W glwt + 2\gamma(t + w)\cos\theta - \rho_L gtwh \qquad (3.3)$$

where γ is the surface tension of the liquid, θ is the contact angle on the solid plate and g is the gravitational constant. The usual procedure is to choose a plate that is completely wetted by the liquid (i.e., $\theta = 0$) and measure the change in F for a stationary plate. The change in force ΔF is then related to the change in surface tension $\Delta\gamma$ by

$$\Delta\gamma = \frac{\Delta F(t + w)}{2} \qquad (3.4)$$

If the plate is thin, so that $t \ll w$

$$\Delta\gamma = \frac{\Delta F}{2w} \qquad (3.5)$$

Some of the precautions needed with the Wilhelmy method are discussed by Middleton et al. (1984).

The Langmuir balance is a differential technique. A clean water surface is separated from the monolayer-covered area by a partition (usually a movable PTFE float connected to a conventional balance) and the force acting on this partition is measured (Gaines, 1966; Petty and Barlow, 1990).

3.7.5 Surface potential measurement

The magnitude of the surface potential ΔV can be measured by placing an electrode coated with an α-emitter, such as polonium or americium, close to the subphase surface (figure 3.14). The resulting air ionization (≈ 1000 times) makes the gap between the probe and the liquid surface ($\approx 5\,\mathrm{mm}$) sufficiently conducting that the potential difference can be measured

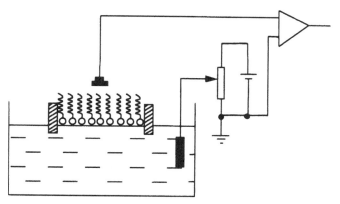

Figure 3.14 Measurement of surface potential.

by means of a high impedance voltmeter. A suitable reference electrode (Ag/AgCl or Pt) is positioned in the subphase.

Alternatively, an electrode in the air is moved with respect to the subphase surface; the resulting change in capacitance then leads to a current flow in the external circuit (Gaines, 1966). The radioactive electrode arrangement is preferred by workers on monolayer systems as this avoids the effect of air flows (due to the vibrating electrode) causing monolayer movement.

3.8 Experimental techniques

Meticulous attention to experimental detail is required for all monolayer and LB film work (Pethica, 1987). Airbound contamination is difficult to avoid completely and precautions such as simple covers or enclosures for the trough are desirable. A substantial reduction in the number of airbound particles near the subphase may be made by locating the trough in a micro-electronics-type clean room. The possibility of contamination originating from the trough operator should not be neglected. Wearing of coats, caps, gloves and face masks (standard procedures in the microelectronics industry) is also advisable.

3.8.1 Surface cleaning

Adequate cleaning of the surface before monolayer spreading is an essential part of obtaining high quality LB layers. With single barrier troughs this can be readily accomplished by sweeping the compression barrier (or a second barrier) over the surface. The usual approach is first to reduce the area to a minimum and then to clean this surface with a glass capillary tube (or a series of tubes) attached to a suitable pump. Surface cleaning may be aided by spreading and removing a monolayer of an appropriate material or

Table 3.1 *Properties of common spreading solvents used for monolayer work.*

Solvent	Boiling point [°C]	Water Solubility at 25 °C $[\text{kg m}^{-3}]$
n-hexane	69	0.01
cyclohexane	81	0.07
chloroform	62	8.0
hexadecane	287	very low

simply by washing the subphase surface with a solvent. The success of the cleaning operation can be determined by monitoring the surface pressure as the area is reduced. The operation may well have to be repeated several times before this is below an acceptable value.

3.8.2 Monolayer spreading

Monolayer-forming materials are applied to the subphase by first dissolving them in a solvent. The necessary properties for such a solvent have been discussed by Gaines (1966). It must dissolve an adequate quantity of the monolayer material (concentrations of $0.1–1\,\text{kg m}^{-2}$ are typical); it must not react chemically with the material or dissolve in the subphase; finally, the solvent must evaporate within a reasonable period so that no trace remains in the condensed monolayer. A solvent that evaporates too quickly may be a problem as this could prevent the accurate determination of the solution concentration. The solvent concentration can also affect the thickness of floating layers formed from compounds that do not form true *mono*layers at the air/ water interface (e.g., unsubstituted molecules, such as the phthalocyanines and fullerenes, Chapter 4).

Some common solvents used in monolayer spreading are listed in table 3.1, along with some of their properties; all these are relatively nonpolar (section 2.2). For the spreading of materials that are not soluble in such solvents, special techniques have to be devised. One approach is to use mixed solvents, which provide the necessary solubility for the monolayer material but do not introduce serious water-solubility problems. Examples include mixtures of the compounds listed in table 3.1 with simple alcohols. Some solvents with quite high boiling points are sometimes used as an aid to spreading and film deposition. One example is hexadecane. This can be thought of as a 'molecular lubricant', which is slowly squeezed out under applied surface pressure. Molecules of such compounds are quite likely to be transferred to the substrate, with the monolayer film, during LB deposition.

The usual way of spreading a monolayer solution is to dispense the mono-layer solution from a microsyringe onto the subphase surface. If the drops are

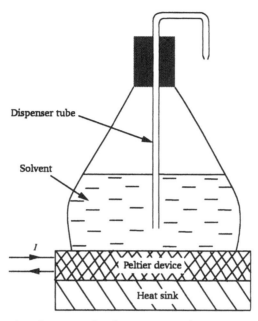

Figure 3.15 Simple system for the automatic dispersion of a monolayer solution. The direction of the current I through the thermoelectric element determines whether the flask is heated or cooled.

allowed to fall near the centre of the subphase surface then any contamination is spread away from the dipping area. It is possible to automate the monolayer spreading process using a motor-driven microsyringe. Figure 3.15 shows an alternative method, a variation of a technique originally suggested by Peterson (1985). The solution is stored in a glass flask that is fixed to a *Peltier* thermoelectric heating element. To spread the solution, a current is passed through the element causing the air in the flask to expand and the solution to be expelled. If the surface pressure is monitored, then the current through the thermoelectric element can be reversed when an adequate quantity of material has been spread. Such a system can be readily incorporated as part of a fully computer-controlled LB deposition apparatus, simplifying the deposition of relatively large numbers of LB layers.

3.8.3 Substrate preparation

Condensed monolayers can be transferred to a variety of surfaces. The adhesion of the first layer to the underlying substrate is particularly critical and will determine the quality of subsequent layers. There are many parameters associated with the substrate surface that can influence deposition. Its exact chemical composition may affect ion exchange in the first layer deposited (section 3.4). The physical structure of the substrate surface (e.g.,

Table 3.2 *Conditions for the LB deposition from the L_2' and S phases of some long-chain fatty acids on pure water subphases.*

(^{a}After Leuthe and Riegler, 1992; ^{b}After Peterson et al., 1988; ^{c}After Song et al.,1993.)

Fatty acid	Formula	Surface pressure [mN m^{-1}]	Temperature [°C]
n-docosanoic (behenic) L_2' phase^a	$C_{21}H_{43}COOH$	25	20
n-docosanoic S phase^a		35	9
22-tricosenoic L_2' phase^b	$C_{22}H_{43}COOH$	30	19
22-tricosenoic S phase^c		45	20

Table 3.3 *Conditions for the deposition on glass slides of monolayers of some cadmium salts of long-chain fatty acids. Transfer from water subphase containing 3×10^{-4} M CdCl$_2$. (Surface pressure − 30 mN m^{-1}.) (After Kuhn et al., 1972.)*

Fatty acid	Formula	Subphase pH	Temperature [°C]
n-octadecanoic (stearic)	$C_{17}H_{35}COOH$	5.4–5.8	18–20
n-nonadecanoic	$C_{18}H_{37}COOH$	5.6–6.0	20
n-eicosanoic (arachidic)	$C_{19}H_{39}COOH$	5.5–6.0	18–22
n-heneicosanoic	$C_{20}H_{41}COOH$	5.6–5.7	25
n-docosanoic (behenic)	$C_{21}H_{43}COOH$	6.0	29

whether it contains gaps or voids) may also be important in determining the quality of the deposited layer (section 3.6). A range of surface preparation techniques for commonly used solid substrates is given by Kuhn et al. (1972) and by Petty and Barlow (1990).

3.8.4   Film deposition

The temperatures and surface pressures must be carefully controlled so that the floating monolayer is well within a single phase region in its phase diagram. Recipes for the successful deposition of simple amphiphilic compounds may be found in many research papers, including the original publications of Langmuir and Blodgett (Blodgett, 1935; Blodgett and Langmuir, 1937). Tables 3.2 and 3.3 list conditions for the successful transfer of long-chain fatty acids and their cadmium salts.

3.9   Post deposition treatments

Following LB deposition it is possible to treat monolayer and multilayer films in several ways. These range from simple processes, such as storage in a desiccator to remove water, to more complex procedures for cross-linking the monomer molecules in an LB array (section 4.6.1). Other common procedures are outlined below.

3.9.1 Skeletonization

Long-chain fatty acid materials are often deposited as a mixture of a fatty acid and a salt. The exact composition of the deposited layer depends on the type of ion in the subphase and the pH of the latter (e.g., equation (2.14)). The free acid may be removed from the film by soaking the LB layers in a suitable solvent. One to five minutes in alcohol or acetone is usually sufficient to remove the free acid from a 50:50 molar ratio cadmium arachidate: arachidic acid LB film. This *skeletonization* process reduces the refractive index (section 7.1) of the multilayer structure (which now consists of about 50% air), making it suitable for use as an antireflection coating for glass.

3.9.2 Removal of outer layers

In air, LB layers of long-chain materials are usually hydrophobic (with the alkyl chains on the outside), while the structures are hydrophilic under the water (figure 3.8). It might be thought that it is relatively easy to produce a polar surface in air by changing the dipping sequence: for example, finishing the LB deposition under water, removing the floating monolayer and withdrawing the multilayer structure through the clean water surface. This usually leads to a hydrophobic film, as the outer layer falls off when the substrate is raised through the clean water surface. Water-wettable fatty acid LB films may be produced by immersing the LB structure in dilute aqueous solutions containing certain polyvalent cations (Gaines, 1966); this treatment probably involves the removal or overturning of molecules in the outer layers of the LB structure. Two other approaches have also been suggested (McLean et al., 1983). Both require the incorporation of the requisite polar group in a polymerizable amphiphilic molecule (e.g., a diacetylene compound, section 4.7.1):

(i) The multilayer is polymerized *under the water surface* (e.g., using ultraviolet light). The floating monomer monolayer is then removed before the substrate with the stabilized polar surface is withdrawn.

(ii) The transfer of the monomer material takes place normally (i.e., no polymerization during deposition) and the deposition is stopped with the substrate submerged. The floating monomer monolayer is then removed, replaced by a simple long-chain fatty acid layer, and the monomer multilayer structure is withdrawn through this. This LB structure will be hydrophobic in air. However, after polymerization the required polar surface may be exposed by removing the outer fatty acid layer in a suitable solvent.

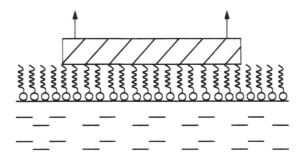

Figure 3.16 Horizontal lifting technique for depositing floating mono-layers. (After Langmuir and Schaefer, 1938.)

3.9.3 Elimination of alkyl chains

Classical LB materials all possess long alkyl chains. These, of course, confer the monolayer-forming ability to the compounds. However, the presence of such chains in the deposited multilayer can reduce the thermal stability of the LB film and can 'dilute' the effect of functional electrical or optical chemical groups in the film. For some materials, it is possible to get the benefit of both worlds. A long-alkyl chain is used for monolayer formation and, after LB deposition, this group is eliminated, usually thermally. This technique has been used to prepare polyamide and polyphenylenevinylene LB films (Nishikata et al., 1988 & 1989).

3.10   Related thin film deposition techniques

The vertical dipping LB process is not the only way to transfer a floating molecular film to a solid substrate or, indeed, to build-up multilayer films. Other methods are based on 'touching' one edge of a hydrophilic substrate with the monolayer-covered subphase or lowering the substrate horizontally so that it contacts the hydrophobic ends of the floating molecules (Gaines, 1966; Petty and Barlow, 1990; Langmuir and Schaefer, 1938). The latter technique is illustrated in figure 3.16. After the solid substrate has made contact with the monolayer, the rest of the monolayer is cleaned away and the plate and deposited film are lifted away. This approach is useful for the transfer of highly rigid monolayers to solid supports. The floating monolayer will also be subject to less disruptive forces than in the LB method.

Chemical means can also be used to build up multilayer organic films (Whitesides and Laibinis, 1990; Ulman, 1991; Tredgold, 1994). A technique pioneered by Sagiv and co-workers, and based on the successive adsorption and reaction of appropriate molecules, is shown in figure 3.17 (Netzer and Sagiv, 1983, Netzer et al., 1983). The molecules adsorb on the solid substrate to form a monomolecular layer. The headgroup reacts with the substrate

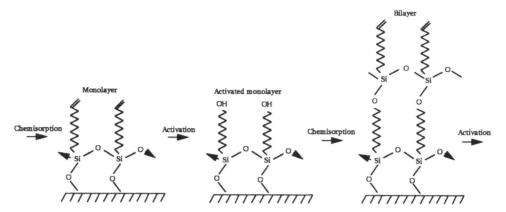

Figure 3.17 Preparation of a chemically attached polymeric multilayer by the Sagiv technique. (After Netzer et al., 1983. Reproduced with permission from Elsevier Science.)

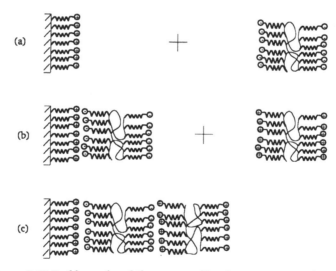

Figure 3.18 Build up of multilayer assemblies by consecutive adsorption of anionic and cationic polyelectrolytes. (After Decher et al., 1992. Reproduced with permission from Elsevier Science.)

to give a permanent chemical attachment and each subsequent layer is chemically attached to the one before in a very similar way to that used in systems for supported synthesis of proteins.

In a similar method, shown in figure 3.18, the ionic attraction between opposite charges is the driving force for the multilayer buildup (Decher

$$L_1 = Cl\,(CH_2)_{11}\,SH$$

$$L_2 = CF_3\,(CF_2)_8\,COOH$$

Figure 3.19 Schematic illustration of the simultaneous formation of two independent self-assembled monolayers. (Reprinted with permission from Whitesides, G. M. and Laibinis, P. E. (1990) Langmuir, *6, 87–96. Copyright 1990 American Chemical Society.)*

et al., 1992 & 1994). In contrast to the Sagiv technique, which requires a reaction yield of 100% to maintain surface functional density in each layer, no covalent bonds need to be formed. A solid substrate with a positively charged planar surface is immersed in the solution containing the anionic polyelectrolyte and a monolayer of the polyanion is adsorbed (figure 3.18(a)). Since the adsorption is carried out at relatively high concentrations of the polyelectrolyte, most ionic groups remain exposed to the interface with the solution and thus the surface charge is reversed. After rinsing in pure water, the substrate is immersed in the solution containing the cationic polyelectrolyte. Again, a monolayer is adsorbed but now the original surface charge is restored (figure 3.18(b)) resulting in the formation of alternating multilayer assemblies of both polymers (figure 3.18(c)).

Cheung et al. (1994) describe a molecular self-assembly technique based on the alternate deposition of a *p*-type doped electrically conductive polymer and a conjugated or nonconjugated polyanion. The delocalized positive charges of the partially doped polymer are used in conjunction with the negatively charged polyanions to assemble the film.

Chemical approaches to the deposition of ultrathin organic films also offer some control over composition and structure in the plane of the surface. An example is shown in figure 3.19 (Whitesides and Laibinis, 1990). A substrate is patterned with gold and aluminium strips using conventional microlithographic techniques. The aluminium oxidizes spontaneously in air and presents aluminium oxide at the solid/vapour interface; the gold remains clean. Two adsorbates, L_1 and L_2, are chosen so that they absorb strongly and selectively on gold and alumina. In the example, L_1 is an alkane-thiol and L_2 a fluorine-labelled carboxylic acid. Exposure of the substrate to a solution containing L_1 and L_2 results in replicating the gold pattern with a self-assembled monolayer derived from the alkane-thiol and the aluminium pattern with a self-assembled monolayer derived from the carboxylic acid.

References

Barraud, A. and Vandevyver, M. (1983) A trough for continuous fabrication of Langmuir–Blodgett films, *Thin Solid Films*, **99**, 221–5

Batey, J. (1983) Electroluminescent MIS structures incorporating Langmuir–Blodgett films, PhD thesis, University of Durham, UK

Bibo, A. M. and Peterson, I. R. (1992) Defect annealing rate in monolayers displaying a smectic-L phase, *Thin Solid Films*, **210/211**, 515–18

Blodgett, K. B. (1934) Monomolecular films of fatty acids on glass, *J. Am. Chem. Soc.*, **56**, 495

Blodgett, K. B. (1935) Films built by depositing successive monomolecular layers on a solid surface, *J. Am. Chem. Soc.*, **57**, 1007–22

Blodgett, K. B. and Langmuir, I. (1937) Built-up films of barium stearate and their optical properties, *Phys. Rev.*, **51**, 964–82

Bonnerot, A., Chollet, P. A., Frisby, H. and Hoclet, M. (1985) Infrared and electron diffraction studies on transient stages in very thin Langmuir–Blodgett films, *Chemical Physics*, **97**, 365–77

Bruinsma, P. J., Spooner, G. J. R., Coleman, L. B., Koren, R., Sturesson, C. and Stroeve, P. (1992) Selectivity of toxic gases over methane and nitrogen in Langmuir–Blodgett films, *Thin Solid Films*, **210/211**, 440–2

Byatt-Smith, J. G. and Malcolm, B. R. (1994) Theoretical and experimental study of the flow of condensed molecular monolayers on a Langmuir trough, *J. Chem. Soc., Faraday Trans.*, **90**, 493–9

Cheung, J. H., Fou, A. F. and Rubner, M. F. (1994) Molecular self-assembly of conducting polymers, *Thin Solid Films*, **244**, 985–9

Decher, G., Hong, J. D. and Schmitt, J. (1992) Buildup of ultrathin multilayer films by a self-assembly process: III. Consecutively alternating adsorption of anionic and cationic polyelectrolytes on charged surfaces, *Thin Solid Films*, **210/211**, 831–5

Decher, G., Lvov, Y. and Schmitt, J. (1994) Proof of multilayer structural organization in self-assembled polycation–polyanion molecular films, *Thin Solid Films*, **244**, 772–7

Gaines, G. L., Jr. (1966) *Insoluble Monolayers at Liquid–Gas Interfaces*, Wiley-Interscience, New York

Grundy, M. J., Musgrove, R. J., Richardson, R. M. and Roser, S. J. (1990) Effect of dipping rate on alternating layer Langmuir–Blodgett film structure, *Langmuir*, **6**, 519–21

Hann, R. A. (1990) Molecular structure and monolayer properties, in *Langmuir–Blodgett Films*, ed. G. G. Roberts, Plenum Press, New York

Honig, E. P. (1973) Molecular constitution of X- and Y-type Langmuir–Blodgett films, *J. Colloid Interface Sci.*, **43**, 66–72

Jones, C. A., Russell, G. J., Petty, M. C. and Roberts, G. G. (1986) A reflection high-energy electron diffraction study of ultra-thin Langmuir–Blodgett films of 22-tricosenoic acid, *Phil. Mag. B*, **54**, L89–L94

Kimura, F., Umemura, J. and Takenaka, T. (1986) FTIR-ATR studies on Langmuir–Blodgett films of stearic acid with 1–9 monolayers, *Langmuir*, **2**, 96–101

Kuhn, H., Möbius, D. and Bücher, H. (1972) Spectroscopy of monolayer assemblies, in *Techniques of Chemistry*, eds. A. Weissberger and B. W. Rossiter, Vol. 1, Part IIIB, pp. 577–702, Wiley, New York

Kumehara, H., Kasuga, T., Watanabe, T. and Miyata, S. (1989) Fabrication of copoly(methacrylic acid–methylacrylate) Langmuir–Blodgett films using a moving-wall-type Langmuir–Blodgett trough, *Thin Solid Films*, **178**, 175–82

Langmuir, I. (1920) The mechanism of the surface phenomenon of flotation, *Trans. Faraday Soc.*, **15**, 62–74

Langmuir, I. and Schaefer, V. J. (1938) Activities of urease and pepsin monolayers, *J. Am. Chem. Soc.*, **60**, 1351–60

Leuthe, A. and Riegler, H. (1992) Thermal behaviour of Langmuir–Blodgett films. II: x-ray and polarized reflection microscopy studies on coexisting polymorphism, thermal annealing and epitaxial layer growth of behenic acid multilayers, *J. Phys. D: Appl. Phys.*, **25**, 1786–97

Lu, Z., Qian, F., Zhu, Y., Yang, X. and Wei, Y. (1994) An automatic Langmuir–Blodgett system based on flowing subphase compression, *Thin Solid Films*, **243**, 371–3

Malcolm, B. R. (1988) The flow of molecular monolayers in relation to the design of Langmuir troughs and the deposition of LB films, *J. Phys. E: Sci. Inst.*, 21, 603–7

Malcolm, B. R. (1989a) Aspects of Langmuir trough design in relation to the study of molecular monolayers of polymers and condensed films, *Thin Solid Films*, 178, 17–25

Malcolm, B. R. (1989b) An apparatus for the study of molecular monolayers compressed at the air–water interface with uniform shear, *Thin Solid Films*, 178, 191–6

McLean, L. R., Durrani, A. A., Whittam, M. A., Johnston, D. S. and Chapman, D. (1983) Preparation of stable polar surfaces using polymerizable long-chain diacetylene molecules, *Thin Solid Films*, 99, 127–31

Middleton, S. R., Iwahashi, M., Pallas, N. R. and Pethica, B. A. (1984) Absolute surface manometry: Thermodynamic fixed points for air–water monolayers of pentadecanoic acid at 25 °C, *Proc. R. Soc. Lond. A*, 396, 143–54

Miller, L. S. and Rhoden, A. L. (1994) A novel Langmuir–Blodgett trough and its applicability to deposition of films of various fluidities, *Thin Solid Films*, 243, 339–41

Miller, L. S. and Stone, P. J. W. (1992) A novel alternating-layer trough, *Thin Solid Films*, 210/211, 19–21

Netzer, L. and Sagiv, J. (1983) A new approach to construction of artificial monolayer assemblies, *J. Am. Chem. Soc.*, 105, 674–6

Netzer, L., Iscovici, R. and Sagiv, J. (1983) Adsorbed monolayers versus Langmuir–Blodgett monolayers – why and how? I: From monolayer to multilayer, by adsorption, *Thin Solid Films*, 99, 235–41

Neugebauer, C. A. (1970) Condensation, nucleation and growth of thin films, in *Handbook of Thin Film Technology*, eds. L. I. Maissel and R. Glang, Chapter 8, McGraw-Hill, New York

Nishikata, Y., Kakimoto, M., Morikawa, A. and Imai, Y. (1988) Preparation and characterization of poly(amide-imide) multilayer films, *Thin Solid Films*, 160, 15–20

Nishikata, Y., Kakimoto, M. and Imai, Y. (1989) Preparation and properties of poly(p-phenylene-vinylene) Langmuir–Blodgett film, *Thin Solid Films*, 179, 191–7

Peterson, I. R. (1985) A fully automated high performance Langmuir–Blodgett trough, *Thin Solid Films*, 134, 135–41

Peterson, I. R. (1990) Langmuir–Blodgett films, *J. Phys. D: Appl. Phys.*, 23, 379–95

Peterson, I. R. (1992) Langmuir–Blodgett films, in *Molecular Electronics*, ed. G. J. Ashwell, pp. 117–206, Research Studies Press, Taunton

Peterson, I. R. and Russell, G. J. (1985) Deposition mechanisms in Langmuir–Blodgett films, *Br. Polym. J.*, 17, 364–7

Peterson, I. R., Russell, G. J., Earls, J. D. and Girling, I. R. (1988) Surface pressure dependence of molecular tilt in Langmuir–Blodgett films of 22-tricosenoic acid, *Thin Solid Films*, 161, 325–41

Pethica, B. A. (1987) Experimental criteria for monolayer studies in relation to the formation of Langmuir–Blodgett multilayers, *Thin Solid Films*, 152, 3–8

Petty, M. C. (1990) Characterization and properties, in *Langmuir–Blodgett Films*, ed. G. G. Roberts, pp. 131–221, Plenum Press, New York

Petty, M. C. (1992) Possible applications for Langmuir–Blodgett films, *Thin Solid Films*, 210/211, 417–26

Petty, M. C. and Barlow, W. A. (1990) Film deposition, in *Langmuir–Blodgett Films*, ed. G. G. Roberts, Plenum Press, New York

Song, Y. P., Petty, M. C. and Yarwood, J. (1993) FTIR studies on the azimuthal distribution of crystallites in 22-tricosenoic acid Langmuir–Blodgett films, *Langmuir*, 9, 543–9

Stephens, J. F. (1972) Mechanisms of formation of multilayers by the Langmuir–Blodgett technique, *J. Colloid Interface Sci.*, 38, 557–66

Suga, K., Iwamoto, Y. and Fujihira, M. (1994) Comparison of LB films obtained with an ordinary trough and a moving-wall type trough by ESR spectroscopy, *Thin Solid Films*, 243, 330–4

Tredgold, R. H. (1994) *Order in Thin Organic Films*, Cambridge University Press, Cambridge

Ulman, A. (1991) *Ultrathin Organic Films*, Academic Press, San Diego

Whitesides, G. M. and Laibinis, P. E. (1990) Wet chemical approaches to the characterization of organic surfaces: self-assembled monolayers, wetting, and the physical-organic chemistry of the solid-liquid interface, *Langmuir*, 6, 87–96

4

Monolayer materials

4.1 Fatty acids and related compounds

A simple long-chain fatty acid such as n-octadecanoic acid (stearic acid) consists of a linear chain (C_nH_{2n+1}) – an *alkyl chain* – terminating in a carboxylic acid group (COOH). The polar acid head confers water solubility while the hydrocarbon chain prevents it (section 2.2). It is the balance between these two opposing forces that results in the formation of an insoluble monolayer at the air/water interface. Any change in the nature of either the alkyl chain or the polar end group will affect the monolayer properties.

The solubility of fatty acids in water decreases as the length of the alkyl chain is increased. To obtain an insoluble monolayer of a nonionized fatty acid (i.e., the situation at sufficiently low pH values), the molecule must contain at least 12 carbon atoms. For example, n-dodecanoic acid (lauric acid – $C_{11}H_{23}COOH$) forms a slightly soluble gaseous monolayer at low temperatures. The addition of two more carbon atoms, to form n-tetradecanoic acid (myristic acid), causes the gas phase to condense at low surface pressures and an expanded monolayer phase to be formed (Stenhagen, 1955). If this monolayer is held at a surface pressure of $10\,mM\,m^{-1}$ and a temperature of $20\,°C$, then the loss in monolayer area due to solubility in the water subphase is $0.1\%\,min^{-1}$. This contrasts with n-octadecanoic acid ($C_{17}H_{35}COOH$ – stearic acid) which shows a decrease in monolayer area of less than $0.001\%\,min^{-1}$ under similar conditions (Gaines, 1966; Hann, 1990). These figures simply reflect the different solubilities of the two long-chain compounds in the water subphase at $20\,°C$: $2.0\,kg/100\,m^3$ and $0.29\,kg/100\,m^3$ for n-tetradecanoic acid and n-octadecanoic acid respectively.

The addition of further carbon atoms also results in the appearance of condensed monolayer phases. Table 4.1 lists some common fatty acids used

Table 4.1 *Long-chain fatty acid compounds used for monolayer studies.*

Structure	Systematic name	Common name	Melting point [°C]
$C_{13}H_{27}COOH$	n-tetradecanoic	myristic	55
$C_{14}H_{29}COOH$	n-pentadecanoic		53
$C_{15}H_{31}COOH$	n-hexadecanoic	palmitic	63
$C_{17}H_{35}COOH$	n-octadecanoic	stearic	71
$C_{19}H_{39}COOH$	n-eicosanoic	arachidic	77
$C_{21}H_{43}COOH$	n-docosanoic	behenic	80
$C_{22}H_{45}COOH$	n-tricosanoic		79

in monolayer work. These tend to be the compounds containing an even number of carbon atoms that occur in nature. The longer chain materials shown in the table are popular with LB workers; high quality multilayer films may be built up readily from n-eicosanoic acid, n-docosanoic acid and n-tricosanoic acid (section 3.8.4).

4.1.1 The headgroup

Long-chain organic compounds terminating in a group other than a carboxylic acid (COOH) may form condensed insoluble monolayers at the air/water interface. The polarity of the headgroup will determine the stability of the layer (Gaines, 1966). The absence of a polar group (i.e., a simple long-chain hydrocarbon) or a weakly polar head (e.g., CH_2I or CH_2Cl) will simply result in drops or lenses on the water surface. On the other hand, if the dipole moment associated with the headgroup is large (e.g., SO_3^-), then the compound becomes too soluble in the aqueous subphase.

The high surface pressure condensed LS, S and CS monolayer states (chapter 2) are found to occur with long-chain amphiphilic compounds possessing a variety of polar ends. This is evidence that these phases are associated with different arrangements of the hydrocarbon chains (Stenhagen, 1955). Examples are shown in table 4.2. At low temperatures, the area per molecule for monolayers of most of these materials is about $0.2\,nm^2$. However, there are some instances where a long straight compound with a polar group at one end does not exhibit the usual high pressure monolayer phases. This may be due to a peculiarly shaped polar head, to interactions between neighbouring polar groups, or simply to the large size of the head group. Particularly interesting examples are the long-chain nitriles (alkyl cyanides, $C_nH_{2n+1}C{\equiv}N$). Although the CN headgroup is small, the limiting area per molecule for these materials is about $0.28\,nm^2$, suggesting that electrical interactions between the polar groups are important in determining the packing in these monolayers (the

Table 4.2 *Different polar head groups used for monolayer and multilayer studies.*

([a]After Hann, 1990. [b]After Gaines, 1966. [c]After Fukuda and Shiozawa, 1980. [d]After Jones, 1987. [e]After Gaines, 1982. [f]After Stenhagen, 1955.)

Class of compound	Chemical formula	monolayer formation	LB film deposition
alcohols	$C_nH_{2n+1}OH$	similar to fatty acids; no dissociation and isotherm independent of pH (2–10) and of dilute salt solutions.	difficult to form LB films.[a,b]
esters	$C_nH_{2n+1}COOR$	ethyl stearate ($C_{17}H_{35}CO_2C_2H_5$) and similar compounds form condensed monolayers with areas per molecule of $\approx 0.2\,\text{nm}^2$.	ethyl stearate may be built into LB multilayers; X-type and Y-type deposition possible.[c]
amides	$C_nH_{2n+1}CONH_2$	condensed isotherms with area per molecule of $\approx 0.2\,\text{nm}^2$.[b]	alternate-layer deposition with a fatty acid.[d]
amines	$C_nH_{2n+1}NH_2$	condensed isotherms for $n > 13$; ionize at low pHs.	docosylamine ($n = 22$) deposits readily.[a,b,e]
nitriles	$C_nH_{2n+1}CN$	limiting area per molecule $\approx 2.8\,\text{nm}^2$ for $n \approx 18$.[f]	

dipole moments of CH_3CN and CH_3COOH are 3.9 D and 1.7 D, respectively).

Some of the compounds shown in table 4.2 also exhibit good LB film deposition under appropriate conditions. Long-chain amines mirror the behaviour of fatty acids. Compounds such as n-docosylamine are protonated on acidic subphases, i.e.,

$$C_nH_{2n+1}NH_2 + H_3O^+ \rightleftharpoons C_nH_{2n+1}NH_3^+ + H_2O \qquad (4.1)$$

The monolayer can be stabilized by negatively charged counterions in the subphase (e.g., SO_4^{2-}). This is analogous to the addition of divalent cations to fatty acid subphases.

Alternate-layer LB deposition (section 3.1) offers a means of building up multilayers of compounds that do not readily transfer as LB films by themselves. The alternation of a long-chain amide with a fatty acid, noted in table 4.2, is just one example. Occasionally the alternate layers deposit more readily than the monolayers of the separate components. For example, when a long-chain fatty acid (e.g., n-tricosanoic acid) is alternated with a long-chain amine (e.g., n-docosylamine), the deposition is facilitated by the transfer of a proton from the acid to the amine head group to form a salt, as shown in figure 4.1.

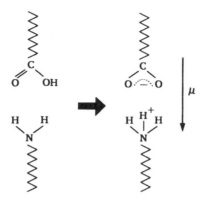

Figure 4.1 Proton transfer in long-chain acid/amine alternate-layer LB film.

4.1.2 The alkyl chain

A simple modification to the alkyl group in a long-chain fatty acid is to replace some, or all, of the hydrogen atoms with fluorine. Since a fluorocarbon chain $(CF_2)_n$ is more hydrophobic than a hydrocarbon chain, it is expected that shorter chains will be needed to confer monolayer-forming properties on a particular polar headgroup. This is found to be the case. Monolayers of many amphiphiles containing fluorocarbon chains are more stable than those formed from their hydrocarbon counterparts (Elbert et al., 1984). The totally fluorinated molecule $C_{10}F_{21}COOH$ and the partially fluorinated species $C_8F_{17}(CH_2)_nCOOH$, with $n = 2$, 4, 6, form stable monolayers (Hann, 1990). The latter material may also be readily transferred to a solid substrate using the LB process. This is significant, as a disadvantage of the LB deposition method (compared to thermal evaporation or spin-coating) is that the alkyl chain is largely redundant in multilayers designed for electronic or electrooptic applications. The presence of these chains not only 'dilutes' the effect of, say, nonlinear dye groups, but also provides a highly insulating region in an LB film intended as a good electrical conductor.

The hydrocarbon chain (C_nH_{2n+1}) in the compounds discussed so far has been *saturated*. This term means that the carbon skeleton is 'saturated' with hydrogen, i.e., besides its bonds with other carbons, each carbon bonds to enough hydrogens to satisfy its valency of four. In saturated hydrocarbon chains, there are only single bonds. If an alkyl chain includes one or more carbon–carbon double or triple bonds, it is referred to as *unsaturated*. The double bond is stronger than a single bond, and also the C=C bond (a *vinyl* group) is much more reactive than the C–C bond. A double bond introduces a constraint because the two parts of a molecule liked by such a bond cannot rotate about it and the bond may disrupt the ordering of the chain. A single *trans*-double bond (appendix A) in the alkyl chain does not produce as much disruption as the

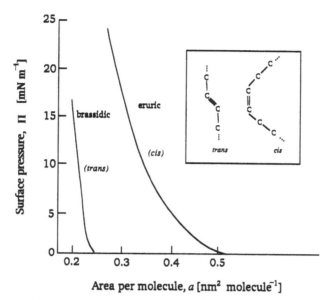

Figure 4.2 Surface pressure versus area isotherms at 21 °C and on 10^{-3} M HCl for eruric and brassidic acids ($C_8H_{17}CH=CHC_{11}H_{22}COOH$). The cis- and trans-double bonds of the two compounds are shown (After Gaines, 1966. Reproduced with permission from G. L. Gaines.)

presence of a *cis*-bond. This can be demonstrated readily using molecular models and is reflected in the melting points of the materials (Hann, 1990). The introduction of *trans*- and *cis*-double bonds into n-docosanoic acid (behenic acid) provides a good example: the saturated fatty acid melts at 80 °C while the melting point of *trans*-13-docosenoic acid (brassidic acid) is 62 °C and that of *cis*-13-docosenoic acid (eruric acid) is further reduced to 34 °C. Figure 4.2 contrasts the surface pressure versus area isotherms for these two unsaturated fatty acids (Gaines, 1966). Clearly, the compound containing the *cis*-bond has a more expanded isotherm than *trans*-material.

A widely studied compound containing a single double bond is 22-tricosenoic acid ($C_{22}H_{43}COOH$), figure 4.3. This molecule has a single C=C bond at the end of the hydrocarbon chain, minimizing the disruption in the packing. The limiting area per molecule for a monolayer of this material is very similar to that for unsaturated long-chain fatty acids. Monolayers may be deposited over a range of surface pressures and temperatures, corresponding to the different monolayer states (chapter 3, table 3.2).

4.2 Simple substituted aromatic compounds

The amphiphilic LB materials discussed to this point may be classified as *aliphatic* compounds. Another major class of organic compounds is called

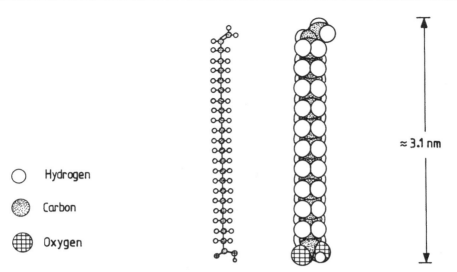

Hydrogen

Carbon

Oxygen

≈ 3.1 nm

Figure 4.3 Chemical structure and space-filling model of 22-tricosenoic acid, showing the position of the C=C bond at the extremity of the alkyl chain.

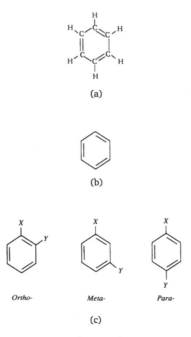

(a)

(b)

Ortho-   Meta-   Para-

(c)

Figure 4.4 (a) Benzene ring. (b) Usual representation of a benzene ring. (c) Isomers of substituted benzene derivatives.

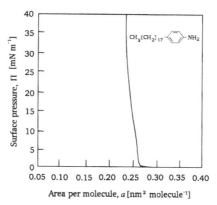

Figure 4.5 Pressure versus area isotherm for 4-dioctadecylaniline. (After Jones, 1987.)

aromatic. The classic aromatic compound is benzene, C_6H_6, with a single six membered ring, figure 4.4(a). The molecule is often depicted as an alternating arrangement of single and double bonds, as shown in figure 4.4(b). However, the bonds between the carbon atoms are all equivalent, with a C–C bond distance (0.14 nm) intermediate between that of a single bond (0.154 nm) and a double bond (0.133 nm). The molecule is sometimes represented by a dotted circle within a hexagon. The *p*-orbitals of the carbon atoms overlap to produce six p molecular orbitals that extend around the benzene ring (appendix A).

Substitution of other chemical groups onto the benzene nucleus allows LB films based on cyclic aromatics to be built up (Hann, 1990). For monolayer formation, one of these is generally a long hydrocarbon chain. The addition of a polar group (e.g., COOH, OH, NH_2) can then provide the amphiphilic requirement for monolayer formation. For such disubstituted benzene derivatives, three possible isomers, depending on the relative positions of the substitutions, are possible. These are distinguished using the prefixes *ortho-, meta-* and *para-* (figure 4.4(c)). The pressure versus area isotherm for a simple *para*-substituted benzene derivative, 4-dioctadecylaniline, is shown in figure 4.5 (Jones, 1987). The limiting area per molecule is about 0.24 nm², significantly larger than the value of 0.19 nm² associated with the close-packing of saturated hydrocarbon chains. This is generally the case for other long-chain amphiphiles containing a benzene ring (Aveyard et al., 1992).

Amphiphilic derivatives of aromatic compounds in which two or more benzene rings are fused or superimposed together at *ortho* positions may also be built up into multilayer films; examples are anthracene (three rings) and pyrene (four rings) (Hann, 1990).

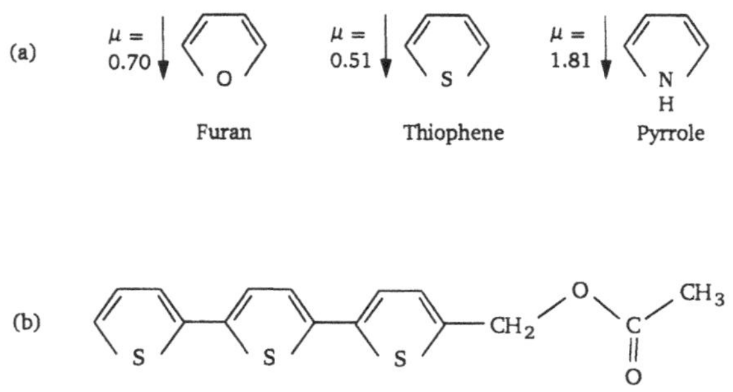

(a)

$\mu = 0.70$ Furan

$\mu = 0.51$ Thiophene

$\mu = 1.81$ Pyrrole

(b)

Figure 4.6 (a) Aromatic heterocyclic compounds. (b) Substituted terthiophene derivative. (After Nakahara et al., 1988.)

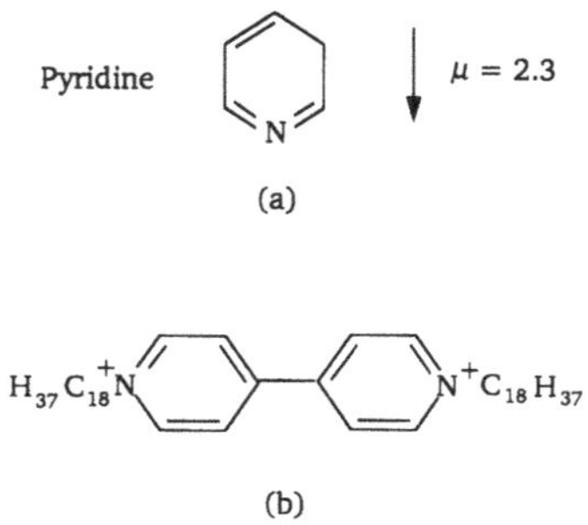

Pyridine $\mu = 2.3$

(a)

(b)

Figure 4.7 (a) Pyridine. (b) Long-chain pyridine derivative for LB film deposition. (After Möbius, 1978.)

Rings containing other atoms besides carbon are called *heterocycles*. The common aromatic heterocycles based on five-membered rings, furan, pyrrole and thiophene, are shown in figure 4.6(a). In these compounds, the dipole moment associated with the π-electron system opposes that due to the σ-bonds (appendix A). In pyrrole, the π dipole moment is larger than the σ moment so that the direction of the net moment μ is opposite to that in the furan and thiophene. Relatively short-chain derivatives of polymers (oligomers) of thiophene form stable layers at the air/water interface at low temperatures, figure 4.6(b) (Nakahara et al., 1988).

The replacement of a carbon atom in a benzene ring by nitrogen renders the molecule much more hydrophilic. Figure 4.7(a) shows the chemical structure for pyridine. The nonbonding pair of electrons (the lone pair – appendix B) on the nitrogen is located in an sp^2 hybrid orbital which is perpendicular to the π-bonds. This ring system in pyridine is relatively electron deficient and the surplus negative charge on the nitrogen gives rise to the large dipole moment of the molecule (2.3 D). In contrast to the five-membered ring systems discussed above, the π moment in pyridine is in the same direction as the σ moment. Figure 4.7(b) shows the structure of

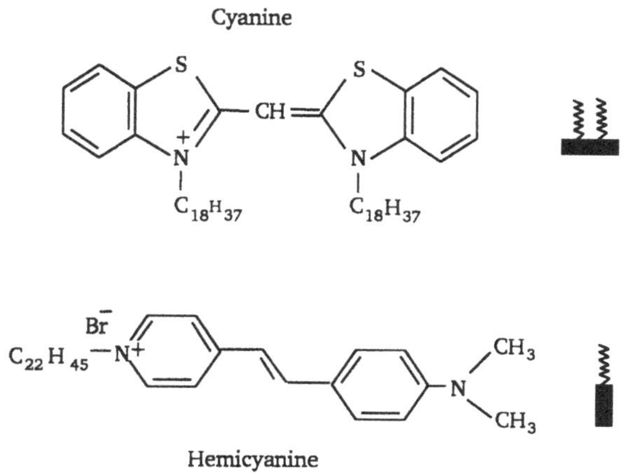

Cyanine

Hemicyanine

Figure 4.8 Amphiphilic cyanine and hemicyanine dye derivatives. (After Kuhn H., Möbius, D. and Bücher, H. (1972) in Techniques of Chemistry, *Vol. 1, Pt. 3B, eds. A. Weissberger and B. W. Rossiter. Copyright © John Wiley & Sons, Inc. Reprinted by permission of John Wiley & Sons, Inc.; Neal et al., 1986.)*

a monolayer-forming compound based on two pyridine rings (Möbius, 1978).

Multilayer LB films may be successfully built up from many of the amphiphilic derivatives discussed in this section. However, the transfer ratios are often less than unity and the film quality usually falls short of that found for films of fatty acids and fatty acid salts. In such cases improvements in deposition may be achieved by mixing with a small quantity (10–20%) of a suitable fatty acid.

4.3 Dyes

Benzene has a $\pi \rightarrow \pi^*$ transition (appendix B) centred at 255 nm, in the ultraviolet part of the electromagnetic spectrum. The addition of a C=C double bond or a further aromatic ring results in an extension to the conjugated π-electron system. *The longer the conjugation, the longer the wavelength of the absorption band.* Compounds with absorption maxima at visible wavelengths generally possess at least two benzene or other aromatic rings linked by a conjugated pathway (based on carbon or nitrogen atoms). Figure 4.8 shows two different approaches that have been used to produce amphiphilic derivatives of such dyes. The cyanine compound was one of those originally used by Kuhn et al. (1972) for energy transfer studies, while the hemicyanine derivative has been synthesized for its nonlinear optical properties (Neal et al., 1986). In the former case, two alkyl chains are substituted onto the dye chromophore. This

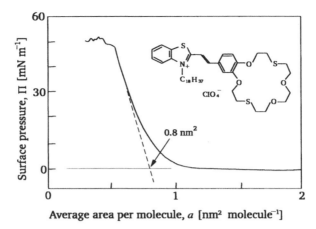

Figure 4.9 Surface pressure versus area isotherm for the amphiphilic benzothiazolium styryl dye shown. (After Lednev and Petty, 1994a.)

results in the chromophore unit lying flat on the subphase surface and a relatively large limiting area per molecule ($\approx 0.5\,\mathrm{nm}^2$) is observed. The transition moment (section B.6) of the main absorption of these dyes is parallel to the long axis of the chromophore. Therefore, if the orientation of the dye molecules is preserved on the substrate surface, s-polarized radiation at the absorption wavelength will be more strongly absorbed than p-polarized radiation, which is observed in practice (Kuhn et al., 1972).

The LB deposition process is likely to orient the chromophores in the hemicyanine derivative so that its long axis is perpendicular to the substrate surface. This is confirmed by the pressure versus area isotherm, which shows a limiting area per molecule in the condensed region of about $0.3\,\mathrm{nm}^2$ (Neal, 1987).

Chromoionophores are more sophisticated versions of simple dyes. They comprise two functionally different chemical groups: an *ionophore*, recognizing specific ions and a chromophore, transducing the chemical information produced by the ionophore–ion interaction into an optical signal. The complexation reaction is influenced by the conformation state (*cis-* or *trans-*) of the chromophore to which it is attached. Figure 4.9 shows an example of an amphiphilic derivative of such a compound (Lednev and Petty, 1994a). High quality multilayers of this compound may be readily built up on solid supports. The optical absorption of the films may then be used to detect reversibly small concentrations of Ag^+ ions in solution (Lednev and Petty, 1994b & c). Such LB films also exhibit partially reversible photochemical behaviour (Lednev and Petty, 1994d).

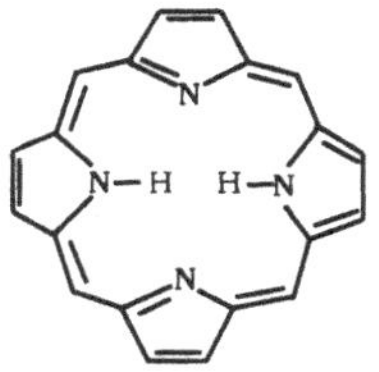

(a)

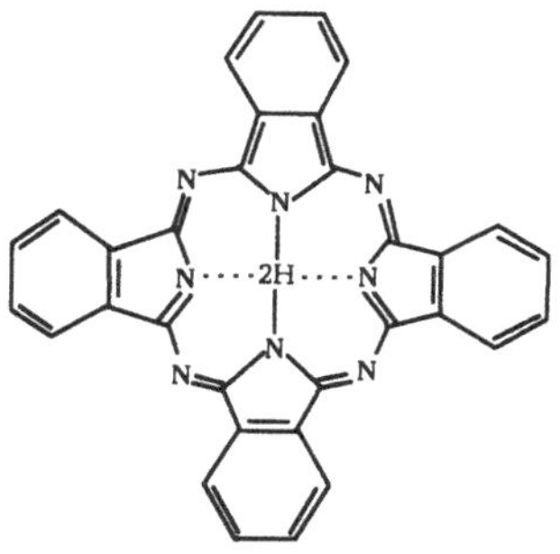

(b)

Figure 4.10 (a) Porphine. (b) Metal-free phthalocyanine.

4.4 Porphyrins and phthalocyanines

An especially important group of compounds containing a ring with alternating double and single bonds is the *porphyrins*. These compounds are ubiquitous in nature and may be regarded as derivatives of the heterocycle *porphine*, figure 4.10(a). The various derivatives differ in the chemical groups attached to the outside of the ring and the nature of the central atom. Some biologically relevant materials (e.g., chlorophyll *a*) already possess the necessary amphiphilic nature to exhibit condensed monolayers at the air/water interface. Such compounds are discussed in section 4.8.2. Long hydrocarbon chains can be attached to the periphery of the porphyrin rings to help monolayer formation and LB film deposition (Hann, 1990).

Related materials are the *phthalocyanines* (Ferraro and Williams, 1987). These compounds are highly coloured, stable at temperatures up to a few hundred degrees Celsius and exhibit a high resistance to chemicals. The parent compound, metal-free phthalocyanine, is illustrated in figure 4.10(b). The substitution of the central hydrogens by metal atoms (e.g., Cu, Ni, Zn) results in changes in the electrical and optical properties of the material. Electrical conductivities range from 10^{-4} to $10^{-1}\,\Omega^{-1}\,m^{-1}$ at room temperature and semiconducting behaviour is normally observed as the temperature is changed (i.e., the conductivity decreases with decreasing temperature with a thermal activation energy of the order of 1 eV); higher conductivities can be achieved by doping.

Using a similar approach to that for the porphyrins, phthalocyanines may be substituted with long alkyl chains around the outside of the conjugated ring system (Hann, 1990). However, it is also possible to form condensed layers at the air/water interface by using derivatives that do not possess the amphiphilic character associated with the classic LB materials. The areas per molecule of many of these compounds show that the floating films are generally *more than one molecule in thickness*. The pressure versus area isotherms are also very dependent on the spreading solvent used, its concentration and the amount initially spread. Z-type deposition is often noted for these unsubstituted compounds and the molecular order in the transferred film is usually much poorer than that associated with long-chain fatty acid type materials.

4.5 Fullerenes

The synthesis of macroscopic quantities of the fullerene C_{60} has led to an extensive range of studies on this molecular allotrope of carbon. The fullerenes C_{60} and C_{70} form condensed layers at the air/water interface (Obeng and Bard, 1991; Williams et al., 1993a; Maliszewskyj et al., 1993). The thickness of the floating layers depends critically on the initial spreading conditions (concentration and quantity of spreading solvent). It is possible to

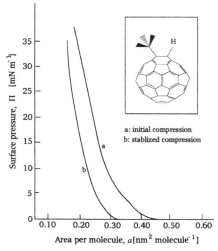

Figure 4.11 Pressure versus area isotherm for 1-t-butyl-9-hydrofullerene-60. The molecular structure is inset. (After Williams et al., 1993b. Reproduced with permission from Elsevier Science.)

use the LB technique to transfer floating films of either C_{60} or C_{70} to solid supports. However, deposition ratios significantly less than unity are observed. A considerable improvement can be made by mixing the fullerenes with a long-chain fatty acid, such as n-eicosanoic acid.

An alternative approach is to use a substituted compound. Figure 4.11 shows pressure versus area curves for 1-t-butyl-9-hydrofullerene-60 (Williams et al., 1993b); the molecular structure is shown in the inset. The two curves correspond to different compressions of the same floating film. Curve a is for the initial compression while curve b is the stabilized isotherm recorded after several compression and expansion cycles; evidently some molecular reorganization has taken place. The isotherms are quite featureless and similar to those observed for many unsubstituted porphyrins and phthalocyanines. The limiting area per molecule of about $0.2\,nm^2$ shows that the floating film is more than one molecule in thickness (from the dimensions of the fullerene derivative, probably five or six). However, this substituted material exhibits much improved LB deposition over the pure C_{60} and C_{70} layers.

4.6 Charge-transfer complexes

Charge-transfer complexes are important organic conductors (Ferraro and Williams, 1987). These are formed from a variety of molecules, primarily aromatics, which can behave as *electron donors* (D) and *electron acceptors* (A). Complete transfer of an electron from a donor to an acceptor molecule results in a system that is electrically insulating (e.g., the transfer of a valence electron in a Cl atom to an Na atom to form the compound NaCl). However, if the ratio of the number of donor molecules to the number of acceptor molecules differs from 1:1, e.g., the *stoichiometry* is 1:2 or 2:3, *or* if there is *incomplete transfer* of an electron from a donor to an acceptor (say, six electrons in every ten donor atoms are transferred), then partially filled electron energy bands can be formed (appendix A) and electrical conduction is possible.

Well-known donor and acceptor molecules are tetrathiafulvalene (TTF) and tetracyanoquinodimethane (TCNQ) respectively (figure 4.12(a)). A 1:1 stoichiometry TCNQ:TTF salt exhibits a high room temperature conductivity ($5 \times 10^4\,\Omega^{-1}\,m^{-1}$) and metallic behaviour is observed as the temperature is reduced to 54 K (i.e., the conductivity increases with decreasing temperature). The molecules in such compounds are arranged in segregated stacks, in which the donors and acceptors form separate donor stacks ($DDDDD\ldots$) and acceptor stacks ($AAAAA\ldots$). The possibility of imitating this structural arrangement in multilayer LB films has resulted in the synthesis of many amphiphilic derivatives of TCNQ, TTF and related compounds (Bryce and

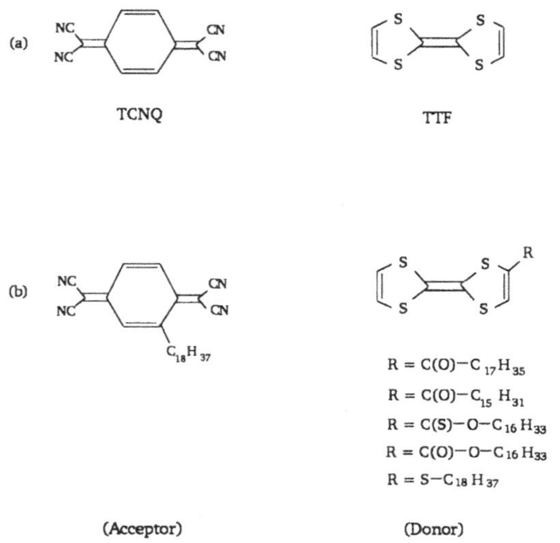

Figure 4.12 (a) The electron acceptor tetracyanoquinodimethane (TCNQ) and the electron donor tetrathiafulvalene (TTF). (b) Amphiphilic derivatives of TCNQ and TTF.

Petty, 1994; Vandevyver, 1992). Some examples are shown in figure 4.12(b). The attachment of the alkyl chain to the side of the TCNQ molecule results in poor packing of the amphiphilic derivative at the air/water interface. Therefore, this compound is usually mixed with a small quantity of a long-chain fatty acid to help LB deposition. The amphiphilic TTF derivatives form rather more condensed layers at the air/water interface and good Y-type deposition is found for many of these compounds. It is also possible to build up alternate layers of the amphiphilic TTF and TCNQ derivatives using the special equipment described in chapter 3, figure 3.11 (Pearson et al., 1989).

4.7 Polymers

Long-chain fatty acid type materials all possess relatively low melting points (table 4.1) and have poor mechanical properties. They are unlikely to find widespread industrial applications. Polymeric materials, on the other hand, are much more robust. There are broadly two different methods available to produce polymeric multilayer structures. First, a monomeric amphiphile that can be deposited using the LB process can be used; 22-tricosenoic acid described in section 4.1.1 is a good example. An alternative is to build up LB films from a polymeric monolayer, i.e., a *preformed* polymer.

4.7.1 Monomer monolayers

Table 4.3 shows some cross-linking groups that have been incorporated into amphiphilic molecules (Hann, 1990). After deposition, the molecules in 22-tricosenoic acid can be cross-linked by exposing the film to ultraviolet radiation. An electron beam may also be used to break the C=C double bonds at the ends of the alkyl chains. Another example of a cross-linkable amphiphile is vinyl stearate, in which the C=C double bond is located close to the polar head group. The polymerization process in the multilayers of such materials normally follows a *free-radical* type solid-state chain reaction (i.e., the intermediates are species with an odd electron on the growing chain). Re-organization of molecules in the polymer film will result in a change in the volume and cracks may appear in the film as it is irradiated. Oxiranes (epoxides) offer a way around this difficulty as there is relatively little volume change during polymerization.

Amphiphilic diacetylenes of the general formula $C_mH_{2m+1}-C{\equiv}C-C{\equiv}C-(CH_2)_n-COOH$ have been widely studied as they can be polymerized conveniently using ultraviolet radiation (Tieke et al., 1983). The quality of the deposited polymer films depends strongly on the molecular structure. A variation in the number n of CH_2 units near the polar end of the molecule has more effect on the packing than a change in the number m of CH_2 groups at the hydrophobic end. Highly ordered films are obtained most easily if a spacer of several CH_2 units is present between the polar head-group and the reactive diacetylene unit; compounds with $m = 12$ and $n = 8$ form good LB layers. Exposure of freshly prepared diacetylene multilayers to ultraviolet light at room temperature results in the formation of a blue coloured film that, upon continuous irradiation, changes to red. These colours are due to the state of conjugation of the polymer backbone. The polymerization proceeds without any major disruption of the multilayer assembly. Dyes may be used to sensitize the photopolymerization process to visible light (Bubeck et al., 1983). The carboxylic acid polar group may be replaced by an alcohol (OH) or an amine (NH_2) termination. In the latter case, the diacetylene compound requires mixing with a long-chain fatty acid as an aid to LB deposition (Tsibouklis et al., 1993).

4.7.2 Preformed polymers

The simplest way of producing a preformed polymer is to take a main chain, having a hydrophilic character due to either polar segments or side groups, and to make it amphiphilic by the attachment of a long alkyl chain. Schematic examples are given in figures 4.13(a) and (b). One of the first preformed polymer systems that was developed was a *co-polymer*, formed by a co-polymerization of maleic anhydride with a range of compounds containing

Table 4.3 *Cross-linking groups that have been incorporated into amphiphilic molecules for LB film deposition.*

Cross-linking group	Examples	Polymerization
C=C	22-tricosenoic acid (C=C bond at end of alkyl chain) vinyl stearate (C=C bond near polar head group)	
C≡C	amphiphilic diacetylenes of general formula R–C≡C–C≡C–X	
O ⟋＼ C — C	amphiphilic oxiranes (epoxides)	

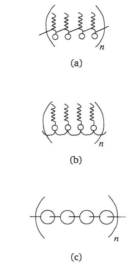

(a)

(b)

(c)

Figure 4.13 Preformed polymer structures.

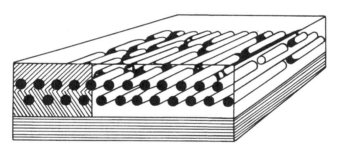

Figure 4.14 Schematic diagram of an LB multilayer formed from rod-like polymers. (After Schaub et al., 1992. Reproduced with permission from Elsevier Science.)

a terminal double bond (Winter et al., 1985; Hann, 1990). Structural investigations of LB films of such materials reveal a uniform layer structure, although the molecules are not necessarily close-packed. The more 'open' structure of the polymer probably helps interdigitation of the side chains.

Figure 4.13(c) shows a rod-like preformed polymer that can be based on the porphyrins or phthalocyanines discussed in section 4.4. These materials undergo self-organization to transferable monomolecular layers at the air/water interface (Schaub et al., 1992). During the transfer process the long axes of the rods are preferentially aligned parallel to the dipping direction, so that oriented multilayers with a nematic-like order are formed, figure 4.14.

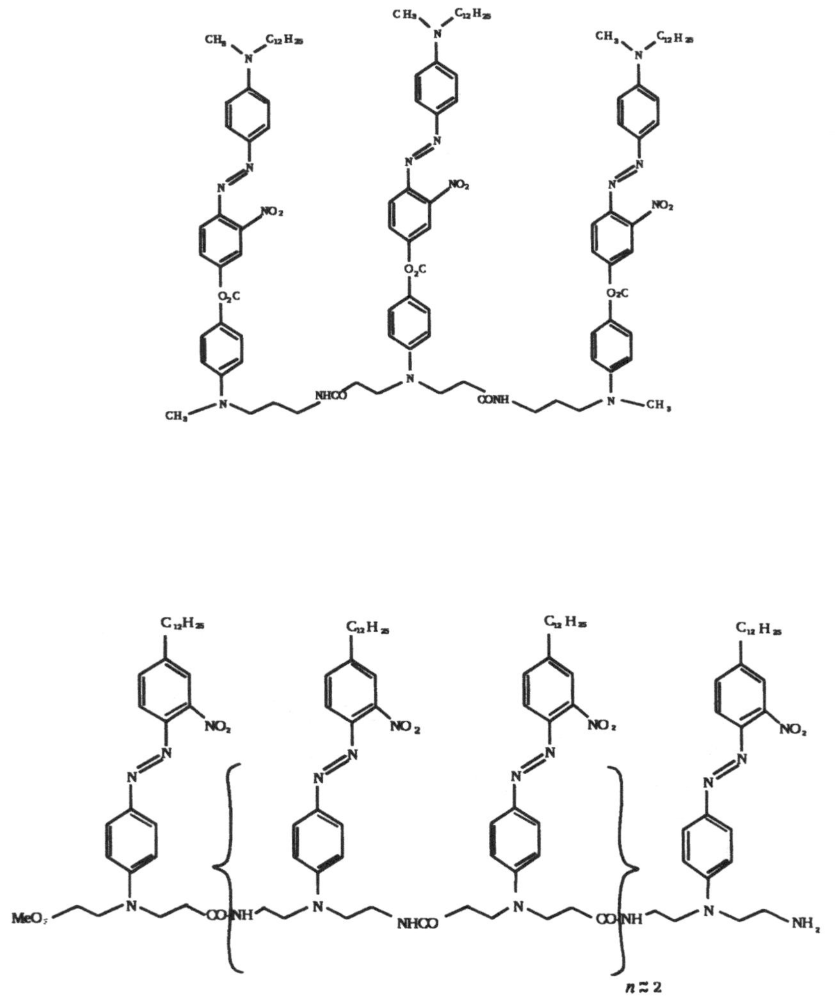

Figure 4.15 Complementary oligomers for alternate-layer deposition. (After Allen at al., 1993.)

It is also possible to incorporate chemical groups with specific electrical or optical properties, e.g., the attachment of chromophores to a polysiloxane spine (Kalita et al., 1992). Figure 4.15 shows a complementary pair of oligomers that has been designed for alternate-layer deposition (Allen et al., 1993). These two compounds can be built up into high quality LB multilayer films (up to several hundred layers in thickness). The optical nonlinearity associated with each monolayer is additive in the multilayer film. Thus, effects such as second-harmonic generation and the linear electrooptic effect may be observed (section 7.7).

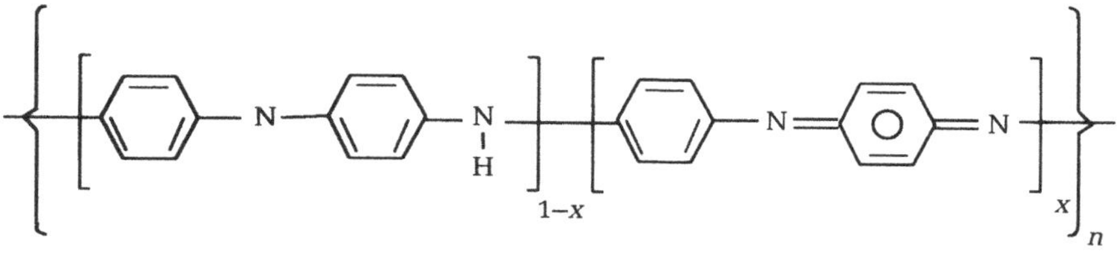

Figure 4.16 Generic formula for polyaniline.

LB films of electrically conductive polymers, e.g., based on repeat units of pyrrole, thiophene or aniline, have attracted interest because of their possible commercial applications. The use of amphiphilic derivatives of thiophene oligomers has already been noted in section 4.2. It is possible to build up multilayers of some unsubstituted conductive polymers. Figure 4.16 shows an example, polyaniline (Agbor et al., 1993). This material is known to possess three reversible oxidation states, each with a distinct backbone structure composed of different proportions of *quinoid* and *phenylene* rings; the figure represents a generic formula for this polymer. To form a multilayer film, polyaniline, in its emeraldine base form ($x = 0.5$ in figure 4.16), is first mixed with a small quantity of acetic acid; this improves spreading on the subphase surface. The mixture is then dissolved in a solvent of chloroform and 1-methyl-2-pyrrolidinone. The resulting isotherm is reasonably condensed, but an area per emeraldine repeat unit of $0.20\,\text{nm}^2$ (at a surface pressure of $30\,\text{mM}\,\text{m}^{-1}$ and temperature of $20\,^{\circ}\text{C}$) shows that the floating layer is not a true monolayer. Nevertheless, Z-type LB films, up to 50 layers in thickness may be built up on solid supports. There are other examples (e.g., thiophene oligomers – Paloheimo et al., 1992) in which the conductive polymers must be mixed with a long-chain fatty acid to obtain reasonable LB film deposition.

A rather different approach is taken by using a *precursor* route to the preparation of LB multilayers of poly(p-phenylenevinylene) (Era et al., 1988; Nishikata et al., 1989). The method consists of three steps: (a) preparation of monolayer films of the amphiphilic precursor polymers at the air/water interface; (b) deposition of the monolayer films onto a suitable substrate, resulting in a multilayer structure of the preformed polymer; and (c) conversion of these films into layers of the target polymers by appropriate heat treatment to remove the long alkyl chains. This technique has also been adopted for the formation of polyamide LB layers (section 3.9.3).

4.8 Biological compounds

Monolayer and multilayer films bear a striking resemblance to the naturally occurring biological membrane, as illustrated in figure 4.17. The basis of this

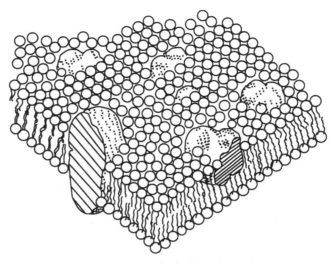

Figure 4.17 Structure of a biological membrane.

structure is a bilayer of amphiphilic phospholipid molecules. *Proteins*, shown as large molecules in the figure, are partially embedded in and partially protruding from this layer. Many components of cell membranes form condensed monolayers at the air/water interface and some may also be built up into multilayer films (Gaines, 1966; Swart, 1990). This has led to interest in the development of LB biomimetic systems, in particular biological sensors (Moriizumi, 1988; Petty, 1991).

4.8.1 Phospholipids

Phospholipids are characterized by two hydrocarbon chains per molecule (not necessarily of the same length). In the most common lipids, the fatty alkyl chains are usually unbranched and one of the two is often unsaturated. The double chains are essential to the correct solid geometry that allows lipids to form membranes while the unsaturated chain helps to maintain the lipid fluidity. Figure 4.18 shows 1,2-dipalmitoyl-*sn*-glycero-3-phosphatidic acid (DPPA), a common phospholipid. The polar end of the molecule is a phosphate group. This is joined to a glycerol unit which, in turn, is attached to the two palmitic acid chains by ester linkages. All phospholipids have a negative charge on the phosphate group at pH 7.0. DPPA forms a condensed layer at the air/water interface, with a limiting area per molecule of approximately 0.4 nm^2 per molecule, as expected from the two hydrocarbon chains. Multilayer films (Y-type deposition) may be built up from either a pure water subphase or one containing divalent cations (Cui et al., 1990; Lukes et al., 1992). In the latter case, a film containing a proportion of salt will be deposited (section 2.6.1).

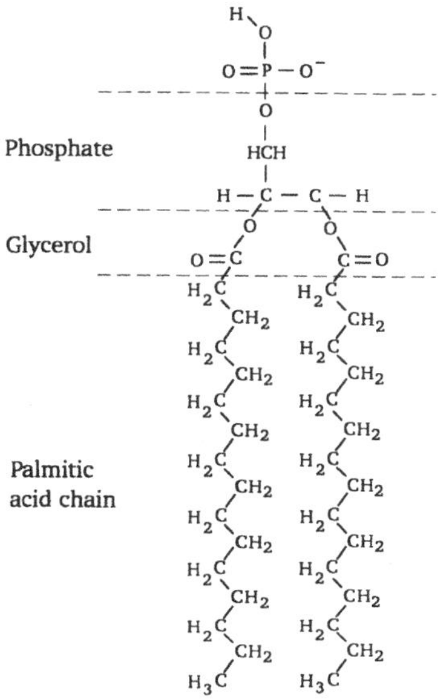

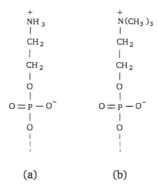

Phosphate

Glycerol

Palmitic
acid chain

Figure 4.18 Structure of 1,2-dipalmitoyl-sn-glycero-3-phosphatidic acid.
The phosphate group will be negatively charged at pH 7.0.

(a)   (b)

Figure 4.19 Polar head groups of (a) phosphatidylethanolamine and (b)
phosphatidylcholine.

Other phospholipids are closely related to the structure shown in figure
4.18. The head groups of two of the most abundant, phosphatidylethanolamine (PE – also called cephalin) and phosphatidylcholine (PC – lecithin),
are shown in figure 4.19. These contain both positive and negative charges
and are called *zwitterionic*. PE and PC materials with appropriate length
chains (e.g., palmitic acid) readily form stable condensed monolayers at
the air/water interface. However, difficulty is often encountered in building
up an LB film that is more than a few (>5) layers in thickness. This may
be associated with the electrostatic nature of the headgroups. A more

successful approach is to deposit mixtures of PC or PE with simple fatty acids or even to build them into alternate-layer structures (Petty et al., 1992). Alternatively, the horizontal dipping method outlined in section 3.10 may be used.

All phospholipids undergo a change of phase from a solid to a liquid crystalline state. This transition is associated with increased disorder and mobility in the fatty alkyl chains. As for simple long-chain fatty acids, the temperature of the transformation depends on the chain length, the number of double bonds and the nature of the head group. Phosphatidylcholine, with two palmitic acid (C_{16}) chains, has a transition temperature of 42 °C. However, if the chain lengths are both reduced by two carbons (to myristic acid chains) then the transition temperature is reduced to 23 °C. The presence of one *cis-* double bond also causes a marked reduction in the transition temperature. Since most of the naturally occurring phospholipids in mammalian cells have at least one such double bond, they will be above the phase transition at body temperature.

The lipids discussed to this point are often called *saponifiable*. This means they can be *hydrolyzed* by heating with alkali to yield soaps of their fatty acid components. Cells also contain nonsaponifiable lipids, examples being the *steroids*. One type of steroid, the *sterols*, is extremely abundant. The best-known compound of this group is cholesterol (cholestene-3-ol), which forms a highly condensed monolayer at the air/water interface with a limiting area per molecule of about 0.40 nm^2 (Gaines, 1966). The isotherm for cholestanol (cholestane-3-ol, a saturated sterol) is very similar; Y-type LB film deposition is possible with this compound (Aveyard et al., 1992).

4.8.2 Pigments

Porphyrin-based biological pigments play important roles in biological energy transport systems. Examples are haematin, the iron porphyrin derived from haemoglobin, and chlorophyll *a*, the green pigment found in higher plants, figure 4.20. Haematin is slightly soluble in water; however, monolayers may be formed using simple derivatives (Gaines, 1966). Chlorophyll *a* forms a stable layer at the air/water interface by itself and may be built up into a multilayer film using the LB technique (Munger et al., 1992; Swart, 1990). These films can form the basis of model membranes for photosynthesis.

The *carotenoids* are photosynthetic pigments that absorb light at wavelengths other than those absorbed by the chlorophylls and serve as supplementary light receptors in a biological membrane. One example, β-carotene, is shown in figure 4.21. In molecular electronics, these compounds are of interest as they contain an unsaturated carbon chain which may act as a *molecular wire* in an appropriate multilayer assembly (Hann, 1990; Oshnishi et al., 1978; Wegmann et al., 1989; Williams et al., 1993c).

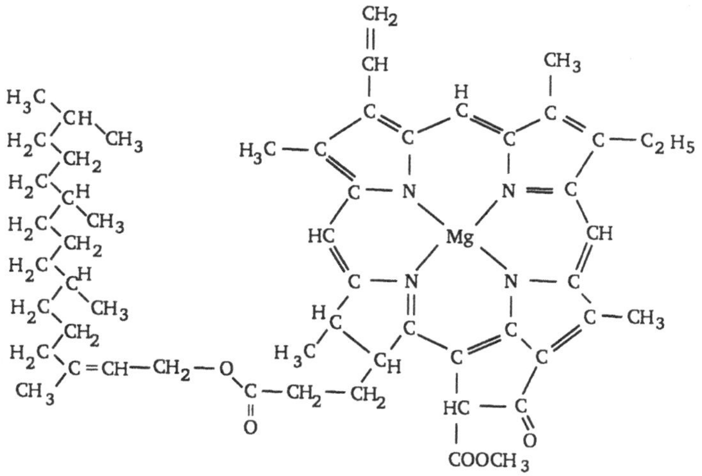

Figure 4.20 Chemical formula of chlorophyll a.

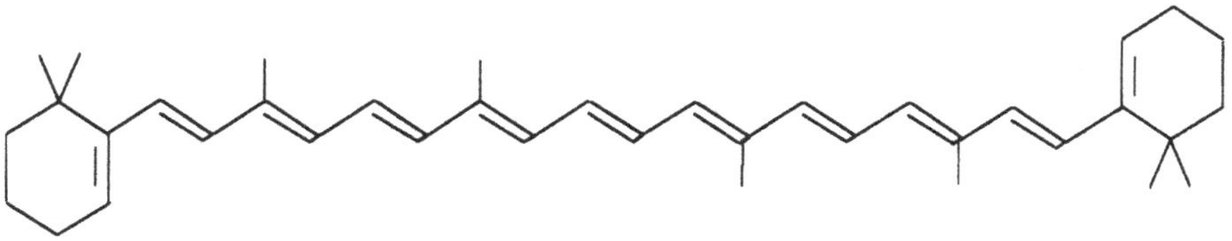

Figure 4.21 Chemical formula of β-carotene.

4.8.3 Peptides and proteins

Proteins are the most abundant macromolecules in most cells and form 50 per cent or more of their dry weight. They are fundamental to all biology since they are the means by which genetic information is expressed. They are also very versatile cell components: for instance, some are *enzymes* (catalyzing chemical reactions), some serve as structural components; and some have *hormonal* activity (regulating biological activity in specific tissues). The building blocks of proteins are the *amino acids* and there are 20 such compounds commonly found in nature. As their name implies, amino acids all contain a basic amino group and an acid carboxyl group. Except for glycine, all these common amino acids are *chiral* molecules, i.e., they possess a carbon atom with four different substituents. Because of the tetrahedral nature of valence bonds on the carbon atom, this leads to two different optical isomers: one will rotate the plane of polarized light to the left and the other to the right.

Amino acids owe their important place in nature to the fact that they may form covalent *amide bonds* between two molecules. Such a linkage is also called a *peptide bond*. As more units are added to the chain, a polymer of any length may be obtained. Such polymers are called *polypeptides* and the individual amino acid units are referred to as *residues*.

Many polypeptides and similar materials can be manipulated at the air/water interface. One such example is the ionophore valinomycin. This is a membrane-active antibiotic capable of selectively complexing with and transporting potassium ions across both biological and synthetic membranes. Complexation is associated with a change in conformation of the ionophore, which facilitates this transport. A diagram of valinomycin, in its complexed state (with potassium), is shown in figure 4.22(a). The molecule is a 12-membered macrocyclic ring of alternating D- and L-amino acids (the small capital letters refer to the configuration of the molecules) and α-hydroxyacids (*depsipeptides*). There are three repeat units, each consisting of L-valine, L-lactic acid, D-valine and D-hydroxyisovaleric acid alternately joined by amide C=O (i.e., part of −CON= group) and ester C=O (i.e., part of −COO− group) linkages.

The valinomycin molecule is disc-shaped and does not possess the cylindrical geometry that is normally associated with long-chain mono-layer-forming materials. In this respect the compound is similar to the unsubstituted phthalocyanines. Figure 4.22(b) shows the pressure versus area plot for this material (Howarth et al., 1988). The expanded region of the isotherm may be accounted for by the large faces of the ionophore being oriented parallel to the water surface. As the monolayer area is reduced, the molecules may re-orient or even become squeezed out of the monolayer, accounting for the plateau in the figure. If the surface pressure is controlled below this constant pressure region, then Y-type multilayers may be built up onto an appropriate support (deposition is helped if the solid substrate is first coated with a few layers of a fatty acid). The ionophore may also be mixed with a long-chain fatty acid or phospholipid; thus, a structure resembling a biological membrane may be fabricated.

Peptides that bridge biological membranes, creating channels for ion transport, may also be incorporated into LB multilayers. An example is gramicidin (Lukes et al., 1992). The helical conformation of this molecule, in which hydrophilic groups occupy the interior and hydrophobic groups the exterior, results in its channel-forming activity. To be active, the channels must necessarily span the entire lipid bilayer (≈5 nm). At present, it is unclear whether the individual gramicidin helices align as a multilayer film is built up.

Proteins are special types of polypeptide. They are large molecules, with molecular weights from 6000 to more than 1 000 000 and can be divided into two classes by their physical characteristics: *globular* and *fibrous*. The structural proteins (examples are protective tissues in hair, skin and nails) tend to be fibrous in nature. The long polypeptide chains are lined up

(a)

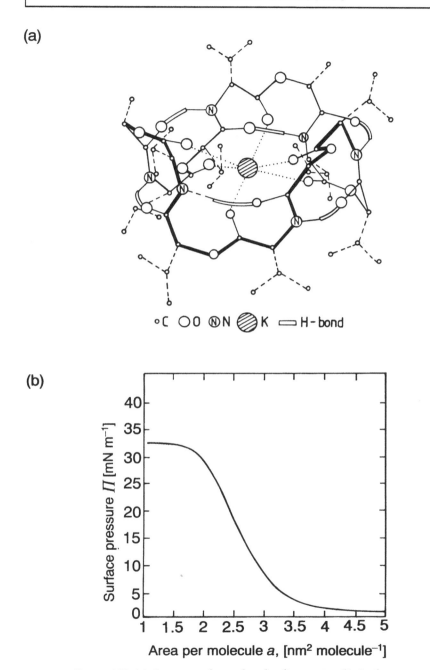

°C ◯O ⓃN ◉K ⊂⊃H-bond

(b)

Figure 4.22 (a) Structure of complexed valinomycin. (b) Surface pressure versus area isotherm for valinomycin on a pure water subphase: pH 5.8, temperature 22 °C. (After Howarth et al., 1988. Reproduced with permission from Elsevier Science.)

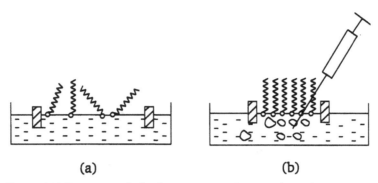

Figure 4.23 Formation of a lipid–protein monolayer at the air/water interface (see text for details).

almost parallel to one another along a single axis; these proteins are usually
insoluble in water.

Globular proteins are generally responsible for regulating the speed of
biochemical reactions (i.e., they act as enzymes) and the transport of various
materials throughout the organisms. The polypeptide chain is folded around
itself in a way that gives the entire molecule a rounded shape. Globular proteins are water soluble and their polar groups tend to be on the outer surface
of the molecule. Such molecules cannot form monolayer films by themselves.
However, they can be incorporated into lipid or fatty acid layers. Figure 4.23
shows how this may be achieved. A floating lipid monolayer is first formed
(figure 4.23(a)) and then a protein-containing solution is injected into the
subphase, just beneath the monolayer (figure 4.23(b)). Alternatively, if a
special multicompartment trough is used, then the floating lipid monolayer
may be transferred (in the condensed state) to a compartment containing
the protein solution (Fromherz, 1975). In both techniques, the protein
molecules are absorbed onto the lipid head groups and a lipid–protein film
is formed at the air/water interface. This layer may then be transferred to a
solid support using the LB method. An important consideration for successful
transfer is the role of electrostatic forces. At low pH the amino acid residues
will become protonated and exist as cations and at sufficiently high pH they
will become negatively charged. However, at some intermediate point the
enzyme will exist in a neutral form; this is called its *isoelectronic point*.
Therefore, the degree of binding to the polar ends of phospholipid or fatty
acid molecules is expected to be dependent on the subphase pH (Zhu et al.,
1989).

The structurally similar proteins streptavidin and avidin have been a model
system for protein binding studies. Each tightly binds biotin at four symmetrically located sites. Streptavidin in an aqueous subphase will bind to a biotin

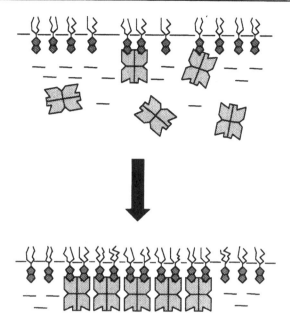

Figure 4.24 Specific binding of streptavidin to a biotin derivatized lipid at the air/water interface.

derivatized lipid at the air/water interface, as shown in figure 4.24. The resulting complex forms two-dimensional crystalline domains (Hoffman et al., 1992).

The three-dimensional arrangement of the polypeptide chains in protein can be quite complex. For instance, the conformation can be helical (as with the gramicidin discussed above) or it can consist of a random coil or a *pleated sheet*. An important question therefore is whether the chains will unfold when the protein is at the air/water interface or during LB film deposition (Swart, 1990). For the time being this question must remain open. However, in one or two specific instances (e.g., glucose oxidase, bacteriorhodopsin) it has been possible to retain biological activity for protein LB films, suggesting that protein unfolding, leading to denaturation, may not be a serious problem (Moriizumi, 1988; Swart, 1990; Li et al., 1992).

References

Agbor, N. E., Petty, M. C., Monkman, A. P. and Harris, M. (1993) Langmuir–Blodgett films of polyaniline, *Synthetic Metals*, **57**, 3789–94

Allen, S., Ryan, T. G., Hutchings, M. G., Ferguson, I., Swart, R., Froggatt, E. S., Burgess, A., Eaglesham, A., Cresswell, J. and Petty, M. C. (1993) Characterization of nonlinear optical Langmuir–Blodgett oligomers, *Proc. OMNO 92 Conference*, eds. G. J. Ashwell and D. Bloor, pp. 50–60, Roy. Soc. Chem., London

Aveyard, R., Binks, B. P., Carr, N., Cross, A. W., Gray, G. W., Kilvert, P. V. A. and Lacey, D. (1992) Effects of monolayer structure on the stability of insoluble monolayers, *Colloids and Surfaces*, 65, 29–42

Bibo, A. M. and Peterson, I. R. (1989) Disclination recombination kinetics in water-surface monolayers of 22-tricosenoic acid, *Thin Solid Films*, 178, 81–92

Bryce, M. R. and Petty, M. C. (1994) *Nature*, 374, 771–6

Bubeck, C., Weiss, K. and Tieke, B. (1983) Sensitized photoreaction of diacetylene multilayers, *Thin Solid Films*, 99, 103–7

Cui, D. F., Howarth, V. A., Petty, M. C., Ancelin, H. and Yarwood, J. (1990) The deposition and characterization of phosphatidic acid Langmuir–Blodgett films, *Thin Solid Films*, 192, 391–6

Elbert, R., Folda, T. and Ringsdorf, H. (1984) Saturated and polymerizable amphiphiles with fluorocarbon chains. Investigation in monolayers and liposomes, *J. Am. Chem. Soc.*, 106, 7687–92

Era, M., Shinozaki, H., Tokito, S., Tsutsui, T. and Saito, S. (1988) Preparation of highly ordered poly(p-phenylenevinylene) thin film by using Langmuir–Blodgett technique, *Chemistry Letters*, 1097–100

Ferraro, J. R. and Williams, J. M. (1987) *Introduction to Synthetic Electrical Conductors*, Academic Press, Orlando

Fromherz, P. (1975) Instrumentation for handling monomolecular films at an air–water interface, *Rev. Sci. Instrum.*, 46, 1380–5

Fukuda, K. and Shiozawa, T. (1980) Conditions for formation and structural characterization of X-type and Y-type multilayers of long-chain esters, *Thin Solid Films*, 68, 55–66

Gaines, G. L., Jr. (1966) *Insoluble Monolayers at Liquid–Gas Interfaces*, Wiley-Interscience, New York

Gaines, G. L., Jr. (1982) Langmuir–Blodgett films of long-chain amines, *Nature*, 198, 544–5

Hann, R. A. (1990) Molecular structure and monolayer properties, in *Langmuir–Blodgett Films*, ed. G. G. Roberts, pp. 17–92, Plenum Press, New York

Hoffman, M., Müller, W., Ringsdorf, H., Rourke, A. M., Rump, E. and Suci, P. A. (1992) Molecular recognition in biotin–streptavidin systems and analogues at the air-water interface, *Thin Solid Films*, 210/211, 780–3

Howarth, V. A., Petty, M. C., Davies, G. H. and Yarwood, J. (1988) The deposition and characterization of multilayers of the ionophore valinomycin, *Thin Solid Films*, 160, 483–9

Jones, C. A. (1987) Pyroelectricity in Langmuir–Blodgett films, PhD thesis, University of Durham, UK

Kalita, N., Cresswell, J. P., Petty, M. C., McRoberts, A., Lacey, D., Gray, G., Goodwin, M. J. and Carr, N. (1992) Nonlinear optical properties of several siloxane polymer Langmuir–Blodgett films, *Optical Materials*, 1, 259–65

Kuhn, H., Möbius, D. and Bücher, H. (1972) Spectroscopy of monolayer assemblies, in *Techniques of Chemistry*, Vol. 1, Pt.. 3B, eds. A. Weissberger and B. W. Rossiter, pp. 577–702, Wiley, New York

Lednev, I. K. and Petty, M. C. (1994a) Langmuir–Blodgett films of a benzothiazolium dye containing a crown ether ring, *Advanced Materials for Optics and Electronics*, 4, 225–32

Lednev, I. K. and Petty, M. C. (1994b) Langmuir–Blodgett films of chromoionophores containing a crown ether ring: complex formation with Ag^+ cations in water, *J. Phys. Chem.*, 98, 9601–5

Lednev, I. K. and Petty, M. C. (1994c) Aggregate formation in Langmuir–Blodgett films of an amphiphilic benzothiazolium styryl chromoionophore, *Langmuir*, 10, 4185–9

Lednev, I. K. and Petty, M. C. (1994d), Photochemistry of an amphiphilic benzothiazolium styryl chromoionophore organized in Langmuir–Blodgett films, *Langmuir*, 10, 4190–4

Li, J. R., Wang, J. P., Chen, T. F., Jiang, L., Hu, K. S., Wang, A. J. and Tan, M. Q. (1992) Photoresponse of Langmuir–Blodgett film containing bacteriorhodopsin, *Thin Solid Films*, 210/211, 760–2

Lukes, P. J., Petty, M. C. and Yarwood, J. (1992) An infrared study of the incorporation of ion channel forming peptides into Langmuir–Blodgett films of phosphatidic acid, *Langmuir*, 8, 3043–50

Maliszewskyj, N. C., Heiney, P. A., Jones, D. R., Strongin, R. M., Cichy, M. A. and Smith, III, A. B. S. (1993) Langmuir films of C_{60}, $C_{60}O$ and $C_{61}H_2$, *Langmuir*, 9, 1439–41

Möbius, D. (1978) Designed monolayer assemblies, *Ber. Bunsenges. Phys. Chem.*, 82, 848–58

Moriizumi, T. (1988) Langmuir–Blodgett films as chemical sensors, *Thin Solid Films*, 160, 413–29

Munger, G., Leblanc, R. M., Zelent, B., Volkov, A. G., Gugeshashvili, M. I., Gallant, J., Tajmir-Riahi, H.-A. and Aghion., J. (1992) Characterization of monolayers and Langmuir–Blodgett films of dry and wet chlorophyll *a*, *Thin Solid Films*, **210/211**, 739–42

Nakahara, H., Nakayama, J., Hoshino, M. and Fukuda, K. (1988) Langmuir–Blodgett films of oligo- and polythiophenes with well-defined structures, *Thin Solid Films*, **160**, 87–97

Neal, D. B. (1987) Langmuir–Blodgett films for nonlinear optics, PhD thesis, University of Durham, UK

Neal, D. B., Petty, M. C., Roberts, G. G., Ahmad, M. M., Feast, W. J., Girling, I. R., Cade, N. A., Kolinsky, P. V. and Peterson, I. R. (1986) Langmuir–Blodgett films for nonlinear optics, *Proc. IEEE Int. Symp. Applications of Ferroelectrics, ISAF '86*, 89–92

Nishikata, Y., Kakimoto, M. and Imai, Y. (1989) Preparation and properties of poly(p-phenylene-vinylene) Langmuir–Blodgett film, *Thin Solid Films*, **179**, 191–7

Obeng, Y. S. and Bard, A. J. (1991) Langmuir films of C_{60} at the air–water interface, *J. Am. Chem. Soc.*, **113**, 6279–80

Oshnishi, T., Hatakeyama, M., Yamamoto, N. and Tsubomura, H. (1978) Electrical and spectroscopic investigations of molecular layers of fatty acids including carotene, *Bull. Chem. Soc. Japan*, **51**, 1714–16

Paloheimo, J., Stubb, H., Yli-Lahti, P., Dyreklev, P. and Inganäs, O. (1992) Electronic and optical studies with Langmuir–Blodgett transistors, *Thin Solid Films*, **210/211**, 283–6

Pearson, C., Dhindsa, A. S., Bryce, M. R. and Petty, M. C. (1989) Alternate-layer Langmuir–Blodgett films of long-chain TCNQ and TTF derivatives, *Synthetic Metals*, **31**, 275–9

Petty, M. C. (1991) Application of multilayer films to molecular sensors: some examples of bioengineering at the molecular level, *J. Biomed. Eng.*, **13**, 209–14

Petty, M., Tsibouklis, J., Petty, M. C. and Feast, W. J. (1992) Pyroelectric behaviour of synthetic biomembrane structures, *Thin Solid Films*, **210/211**, 320–3

Schaub, M., Mathauer, K., Schwiegk, S., Albouy, P.-A., Wenz., G. and Wegner, G. (1992) Investigation of molecular superstructures of hairy rodlike polymers by X-ray reflection, *Thin Solid Films*, **210/211**, 397–400

Stenhagen, E. (1955) Surface films, in *Determination of Organic Structures by Physical Methods*, eds. E. A. Braude and F. C. Nachod, pp. 325–71, Academic Press, New York

Swart, R. M. (1990) Monolayers and multilayers of biomolecules, in *Langmuir–Blodgett Films*, ed. G. G. Roberts, pp. 273–316, Plenum Press, New York

Tieke, B., Lieser, G. and Weiss, K. (1983) Parameters influencing the polymerization and structure of long-chain diynoic acids in multilayers, *Thin Solid Films*, **99**, 95–102

Tsibouklis, J., Pearson, C., Song, Y., Warren, J., Petty, M., Yarwood, J., Petty, M. C. and Feast, W. J. (1993) Pentacosa-10,12-diynoic acid/henicosa-2,4-diynylamine alternate-layer Langmuir–Blodgett films: synthesis, polymerisation and electrical properties, *J. Mater. Chem.*, **3**, 97–104

Vandevyver, M. (1992) New trends and prospects in conducting Langmuir–Blodgett films, *Thin Solid Films*, **210/211**, 240–5

Wegmann, A., Tieke, B., Pfeiffer, J. and Hilti, B. (1989), Electrical conductivity of novel Langmuir–Blodgett films containing ethyl-β-apo-8'-carotenoate, an amphiphilic carotene, *J. Chem. Soc., Chem. Commun.*, 586–8

Williams, G., Pearson, C., Bryce, M. R. and Petty, M. C. (1993a) Langmuir–Blodgett films of the fullerenes C_{70} and C_{60}, *Synthetic Metals*, **55/56/57**, 2935–60

Williams, G., Soi, A., Hirsch, A., Bryce, M. R. and Petty, M. C. (1993b) Langmuir–Blodgett films of 1-*t*-butyl-9-hydrofullerene-60, *Thin Solid Films*, **230**, 73–7

Williams, G., Bryce, M. R. and Petty, M. C. (1993c) Electrical properties of multilayer films containing a carotene derivative, *Mol. Cryst. Liq. Cryst.*, **229**, 83–90

Winter, C. S., Tredgold, R. H., Vickers, A. J., Khoshdel, E. and Hodge, P. (1985) Langmuir–Blodgett films from pre-formed polymers: derivatives of octadec-1-ene-maleic anhydride copolymers, *Thin Solid Films*, **134**, 49–55

Zhu, D. G., Petty, M. C., Ancelin, H. and Yarwood, J. (1989) On the formation of Langmuir–Blodgett films containing enzymes, *Thin Solid Films*, **176**, 151–6

5

Structure of multilayer films

5.1 Organization in long-chain materials

Long-chain aliphatic materials pack together with their hydrocarbon chains parallel. The simplest scheme is an hexagonal array, with the molecules freely rotating as rigid rods about their long axes. The diameter of the cylinder into which one molecule fits is about 0.48 nm (Kitaigorodskii, 1961). Such a plastic crystalline state (section 1.2), originally known as a rotator phase (Ungar, 1983), may be exhibited by straight chain alkanes and some monolayers just below their melting point. There may even be some similarity with the LS phase in floating monolayer films (section 2.4.3). However, there is still considerable debate about this (Ulman, 1991).

For *infinite* aliphatic molecules, the hydrocarbon chain takes the form of a zig-zag, repeating at 0.254 nm intervals along the chain axis. In the most stable state, all the CH_2 group carbon atoms lie in a plane to give a flat zig-zag (appendix A). Close-packed structures result from the hydrogen atoms in a CH_2 group on one molecule fitting into depressions between hydrogen atoms on adjacent molecules. Different packing arrangements of the C_2H_4 repeat units define the crystallographic nature of the *subcell* or *sublattice*. There are three possible close-packed structures with similar packing densities: *orthorhombic* (R), *monoclinic* (M) and *triclinic* (T) (appendix C) (Kitaigorodskii, 1961). In the T and M arrangements, the zig-zags in the chain are parallel to one another. However, this is not so for the R structure, illustrated in figure 5.1. The dimensions of the resulting subcell, a_0, b_0 and c_0 (with c_0 measured along the long axis of the molecule) are

$$a_0 = 0.740 \text{ nm}; \qquad b_0 = 0.495 \text{ nm}; \qquad c_0 = 0.254 \text{ nm}.$$

As evident from figure 5.1, the R packing has two molecules per unit subcell. Structures based on M and T subcells only possess one molecule per unit cell.

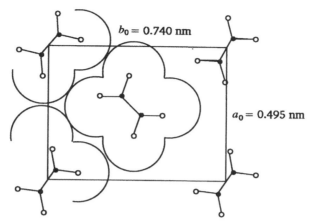

Figure 5.1 Orthorhombic (R) subcell packing for long alkyl chains. View down the long axis of the molecules. ○ hydrogen, ● carbon.

This has important consequences for some infrared bands (section 5.5). If the R subcell is aligned with its c_0 axis perpendicular to the substrate, then the area occupied by each molecule $= 0.5 \ (0.740 \times 0.495) = 0.183 \ nm^2$.

For *finite* length molecules, stereochemical effects associated with the end groups will also be important in determining the packing (and in some LB materials, steric and chemical interactions between the polar heads will *solely* be responsible for the molecular organization). Energetic considerations favour the association of long-chain fatty acid molecules into *dimers*, in which the polar carboxylic acid head groups in one layer are hydrogen bonded to those in the adjacent monolayer. The first and second carbons in the chain after the carboxyl carbon (C_α and C_β) can either be *cis*- or *trans*- (appendix A) with respect to the C=O and C–C bonds. This leads to two possible conformations for the cyclic dimer, depicted in figure 5.2. The diagram also shows the arrangement of a sideways dimer structure, in which the OH group from one fatty acid molecule is hydrogen bonded to the oxygen on an adjacent molecule in the same monolayer.

From the above, it is evident that two levels of organization must be considered for long-chain molecules in layered structures. These are illustrated in figure 5.3. Besides the nature of the subcell, the packing of the individual molecules in the multilayer will define the *main cell*. In the solid state, long-chain fatty acid type compounds exhibit *polymorphism*, i.e., they can exist in a variety of crystallographic states, depending on the packing arrangements of the main and subcells. For example, figure 5.4 shows the structure of the C-form of n-octadecanoic acid (stearic acid). Here the main cell is monoclinic with lattice parameters $a = 0.936 \ nm$; $b = 0.495 \ nm$ and $c = 5.070 \ nm$ (Malta et al., 1971). (Note that b is the same as b_0); the subcell

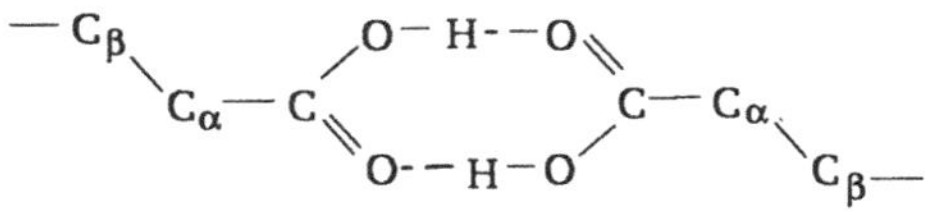

Cyclic *trans* -

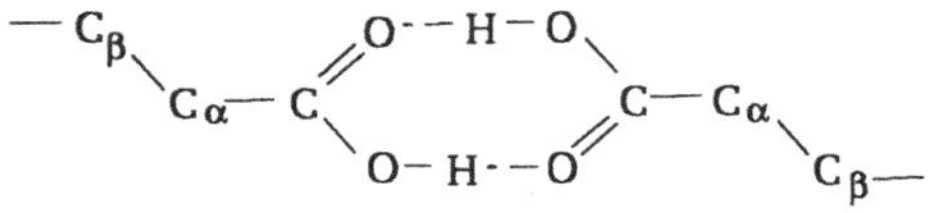

Cyclic *cis* -

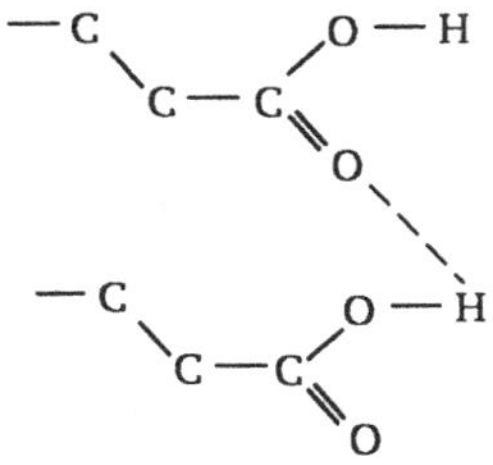

Sideways

Figure 5.2 The cis- *and* trans-*conformations of carboxylic acid cyclic dimers. The structure of a sideways dimer is also given.*

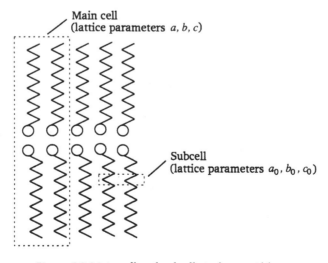

Figure 5.3 Main cell and subcells in fatty acid layers.

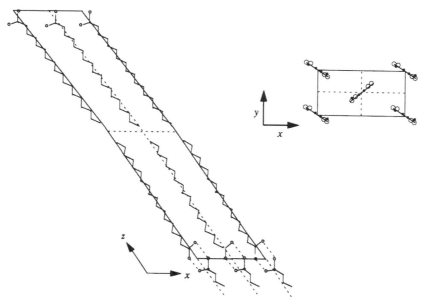

Figure 5.4 Projections on the xz and xy planes of the C-form of n-octa-decanoic acid. (After Malta et al., 1971. Reproduced with permission from the Royal Society of Chemistry.)

is orthorhombic. The projection on the xy plane in the figure reveals that this crystallographic form has two fatty acid dimers (four molecules) per unit cell.

Other main cell structures can result from simple displacements of adjacent molecules along crystallographic axes. In all, there are ten different packing arrangements for long-chain fatty acids depending on the relative displacements of neighbouring chains (Kitaigorodskii, 1961). These can conveniently be distinguished by the *Miller indices* (appendix C) of the interface plane between layers of molecules and the type of subcell symmetry (Peterson and Russell, 1984). For example, the structure shown in figure 5.4 would be designated R(011).

LB films of *fatty acid salts* invariably consist of upright molecules with a R(001) packing (Russell et al., 1984). However, several different structures for *fatty acid* multilayers have been identified, depending on the precise deposition parameters. For 22-tricosenoic acid, the R(001) structure is found over a wide range of dipping conditions (Peterson and Russell, 1984; 1985a & b).

If the LB multilayer structure were a simple crystalline system, the above identification system would allow only a few possible orientations for the molecular chains. In practice, it is found that the tilt angle of the alkyl chains in fatty acid materials (e.g., 22-tricosenoic acid) varies continuously

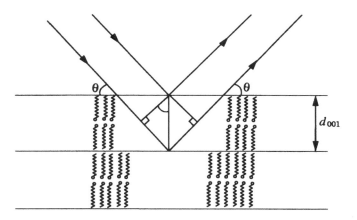

Figure 5.5 Bragg reflection from LB film planes with spacing d_{001}.

with the deposition pressure (section 5.4.3); the greater the pressure, the smaller the angle of tilt from the substrate normal (Peterson et al., 1988; Robinson et al., 1989). The structure of these LB layers of tilted molecules may therefore be quite complex, consisting of regions of crystallinity in which grains are inclined to the substrate with a distribution of tilt azimuths and permeated by holes (Robinson et al., 1989).

The following sections describe the more powerful techniques that can be used to investigate LB film structure and outline some important results.

5.2   X-ray reflection

X-ray diffraction was one of the original methods used to study the structure of fatty acid multilayer films. It led to the discovery that some X-deposited LB films possessed the same structure as Y-type layers (section 3.3).

When electromagnetic energy is incident on the surface of a material, some of it will be reflected *specularly*, so that the angle of incidence is equal to the angle of reflection (well known for visible light, appendix B, equation (B.11)). For crystalline materials, constructive and destructive interference between the radiation reflected from successive crystal planes will occur when the wavelength λ of the incident radiation is of the same order as the lattice spacing (0.1–5 nm), figure 5.5. The condition for maxima in the reflected radiation is provided by *Bragg's law*

$$n\lambda = 2d_{hkl}\sin\theta \tag{5.1}$$

θ is the angle of incidence, conventionally measured from the plane of reflection in X-ray crystallography (in contrast to geometrical optics, appendix B, figure B.3), d_{hkl} is the interplanar spacing (appendix C) and the integer n is known as the *order of the reflection*. Clearly, high-order ($n = 2$, 3, etc.)

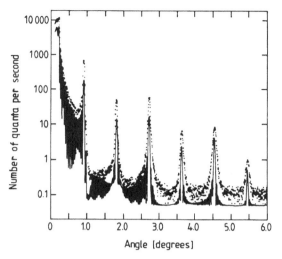

Figure 5.6 X-ray diffraction data from a 43-layer LB film of perdeuterated manganese octadecanoate on a silicon substrate. Experimental values are shown as points. The solid curve is based on calculation and is displaced from the data points. (Reprinted (abstracted with permission) from Nicklow, R. M., Pomerantz, M. & Segmuller, A. (1981) Physical Review B, **23**, *1081–7. Copyright 1981 The American Physical Society.)*

scattering from planes d_{hkl} is indistinguishable from first-order scattering from planes $d_{hkl}/2$, $d_{hkl}/3$, etc.

A small correction is required to equation (5.1) to take into account the fact that the refractive index for X-rays will be slightly less than unity. Bragg's law becomes

$$n\lambda = 2d \sin\left(1 - \frac{\delta}{\sin^2\theta}\right) \tag{5.2}$$

where δ is related to the X-ray refractive index by $n = 1 - \delta$. Typically δ is of the order 10^{-6}. (A refractive index less than unity may be thought to be impossible as it leads to an electromagnetic wave velocity greater than the speed of light (appendix B). However the refractive index is related to the *phase velocity* of the radiation, i.e., the speed at which the crests of the wave travel (Feynman et al., 1963).)

As shown in figure 5.5, the simplest interplanar spacing in an LB multilayer film is d_{001}, which for a fatty acid salt (e.g., cadmium octadecanoate) is approximately 5.0 nm. Using equation (5.1) with an X-ray wavelength of 0.154 nm, gives $\theta \approx 1°$ for the first-order Bragg peak. Therefore, low-angle measurements are necessary for X-ray studies of LB films of long-chain compounds.

Figure 5.6 shows X-ray diffraction data from a 43-layer fatty acid salt film (Nicklow et al., 1981). As predicted, the first-order Bragg reflection is

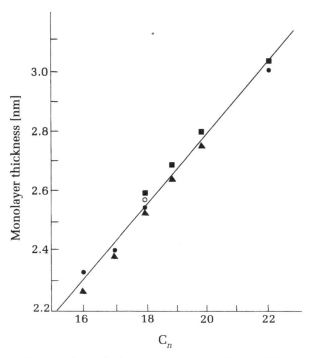

Figure 5.7 Monolayer thickness, obtained from X-ray diffraction experiments, versus number of carbon atoms in the molecule for salts of long-chain fatty acids. ● Srivastava and Verma, 1966; ■ Mann and Kuhn, 1971; ▲ Matsuda et al., 1977. (After Petty, 1990. Reproduced with permission from Plenum Press.)

observed for an angle of incidence close to one degree. In a Y-type LB film, the d_{001} spacing is equal to the distance between the polar planes, i.e., the thickness of two monomolecular layers. The monolayer thicknesses for fatty acid *salts*, obtained from X-ray experiments, are plotted as a function of the number of carbon atoms in the molecule in figure 5.7. The X-ray data are in close agreement with those calculated from the lengths of the molecules, inferring that the hydrocarbon chains in transferred monolayers are oriented almost at right angles to the substrate surface. For Y-type LB films of other amphiphilic materials (including simple fatty acids), the X-ray *d*-spacing is often less than twice the molecular length, suggesting some *tilting* or *interdigitation* (or both!) of the molecules. Unfortunately, the *d*-spacing value alone cannot be used to distinguish between these situations.

X-rays are scattered by their interaction with atomic electrons and interference takes place between X-rays scattered from different parts of an atom. The *scattering power* or *scattering factor* decreases with increasing scattering angle 2θ, resulting in a decrease in the intensity of the Bragg peaks. However, it is evident that the height of the third Bragg reflection in figure 5.6 is greater than that of the second. This is because the intensity of a particular Bragg peak is not only related to the scattering power of each atom in the lattice but also to the position of the atom in the unit cell. In summing the individual waves to give the resultant diffracted beam, both the amplitude and phase of each wave scattered by the individual atoms are important. The intensity of the scattered radiation I_{hkl} from a set of planes 'hkl' may be written (Cheetham, 1988)

$$I_{hkl} = sLpF_{hkl}^2 \qquad (5.3)$$

where s is a scale factor, L is the *Lorentz* (geometrical) correction, p is a polarization correction and F_{hkl} is called the *structure factor*.

The structure factor of the unit cell depends upon the constituent atoms and their individual scattering factors. For a particular d-spacing, the only variable in the cell structure is the nature of the atom. Since F_{hkl} depends directly on the effective number of scattering electrons per atom, a large F_{hkl} value is given by a high atomic number. Thus fatty acid *salt* films, containing heavy metals such as cadmium, barium or lead, are generally superior for X-ray diffraction experiments.

The width of the Bragg peaks at half-height may be used to determine the *correlation radius* R_c according to (Feigin et al., 1989)

$$R_c = \frac{\lambda}{\Delta(2\sin\theta)} \qquad (5.4)$$

The value of R_c characterizes the quality of the packing of the layers on one another and indicates the distance at which an error accumulates in the packing.

For small (<20) numbers of layers, secondary maxima are often observed between the Bragg reflections (Pomerantz and Segmüller, 1980). The intensity and angular position of these depend on the exact number of deposited layers. The phenomenon is analogous to the diffraction of light waves from gratings of only a few slits.

X-ray diffraction can also be used to obtain information about the in-plane structure of LB molecular assemblies. However, since most LB film substrates (e.g., glass slides) will almost completely attenuate an X-ray beam, diffraction in transmission either requires the LB film to be removed from the substrates or special substrate materials to be used. The technique of Prakash et al. (1985) consists of tilting the plane of the multilayer to small angles with respect to the horizontal (with the incident and diffracted X-ray beams in the horizontal plane). This allows the observation of (111) diffraction peaks.

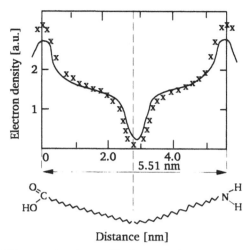

Figure 5.8 Electron density profile for a 60-layer n-tricosanoic acid/1-docosylamine alternate-layer film deposited onto silicon. The crosses and full line represent the results of different iterative procedures. (After Lvov et al., 1989. Reproduced with permission from Taylor and Francis.)

5.2.1 Electron density profile

If the structure factors F_{hkl} for a complete set of X-ray reflections are known, the electron density ρ at any position xyz in the unit cell may be calculated using

$$\rho(xyz) = \frac{1}{V} \sum_h \sum_k \sum_l F_{hkl} \cos 2\pi(hx + ky + lz) \qquad (5.5)$$

where V is the unit cell volume. This is an example of a *Fourier* summation; *the electron density profile is the Fourier transform of the X-ray diffraction pattern.* To calculate $\rho(xyz)$, values of F_{hkl} are needed. Unfortunately, only F_{hkl}^2 can be measured. (The $+$ or $-$ signs of F_{hkl} correspond to phases of $0°$ and $180°$.) It is therefore possible to obtain $|F_{hkl}|$, *but not the sign of the structure factor.* This dilemma is known as the *phase problem* (Tredgold, 1994).

For a set of data containing m structure factors, each of which may be positive or negative, there are 2^m possible combinations of phases; e.g., for 300 reflections there are 2×10^{90} possibilities! Clearly, the solution of the phase problem by trial-and-error methods is not very practicable! A physical criterion is often used to evaluate the electron density profile of multilayer films perpendicular to the substrate. For example, the optimization of a plateau in the electron density profile (representing the region of constant electron density corresponding to the alkyl chain) can be used to simplify these computations in many LB film systems (Belbeoch et al., 1985). Figure 5.8 shows the electron density perpendicular to the substrate for a fatty

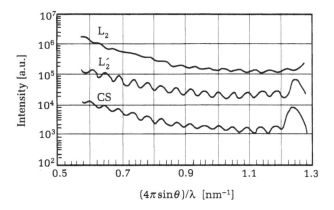

Figure 5.9 Kiessig fringes of n-docosanoic acid multilayers prepared from the L_2, L_2' and CS phases. (After Leuthe and Riegler, 1992. Reproduced with permission from the Institute of Physics.)

acid/fatty amine alternate-layer LB film (Lvov et al., 1989). Regions of high electron density correspond to the polar parts of the molecules. Sometimes, electron density profile calculations can reveal overlap of the molecules in adjacent layers, thus providing complementary information to the d-spacing values (Belbeoch et al., 1985).

5.2.2 Kiessig fringes

The observation of X-ray interferences, first described by Kiessig, is one of the most precise methods to measure the thickness of a thin film (Kapp and Wainfan, 1965). Specular reflection of X-rays by a flat surface is observed at glancing angles close to the critical angle θ_C (appendix B, figure 3.4) for total reflection. For all materials, θ_C is a few tenths of a degree. In a thin film system, each interface of two layers with different refractive indices gives rise to a specularly reflected X-ray beam. With monochromatic X-radiation, interference maxima and minima can be observed near the critical angle. Kiessig showed that it is possible to determine the thickness of the film from the angular spacing of these intensity maxima. If deviation of the X-ray refractive index from unity is neglected, the fringe maxima θ_m are related to the total thickness t by (Tippmann-Krayer et al., 1991)

$$\sin \theta_m = n \frac{\lambda}{2t} \tag{5.6}$$

The film thickness can therefore be obtained from the slope of the $\sin \theta_m$ versus m plot. Figure 5.9 shows examples of Kiessig fringes from n-docosanoic acid multilayers prepared from the L_2, L_2' and CS phases (section 2.4.3) (Leuthe and Riegler, 1992). In this figure the X-ray intensity is plotted as a function of the *scattering vector*, $(4\pi/\lambda)\sin\theta$. The L_2 sample shows the least intense fringes, indicating that the L_2' and CS samples are smoother at the surface. The average bilayer spacing was 4.88 ± 0.08 nm for the L_2 sample, 5.07 ± 0.03 nm for the L_2' layer and 5.07 ± 0.04 nm for the CS film.

X-ray interference phenomena may also be observed in the nonspecularly reflected (scattered) radiation from multilayer films (Kapp and Wainfan, 1965).

5.2.3 Standing X-rays

The positions of particular kinds of atoms may be determined within a layered structure using X-ray standing waves. The principle of the method is similar to that in which the superposition of an incident light wave and its reflection from a mirror produces interference. At a Bragg reflection, there is similarly an intense reflected wave that can interfere with the incident beam. As the angle is varied within the Bragg peak, the positions of the interference minima and maxima above the surface move. An LB multilayer film itself can be used as the diffracting layer. Alternatively, inorganic multilayers such as platinum–carbon or tungsten–silicon can be employed. The intense electric fields in the standing wave field (synchrotron radiation is used) cause electronic excitations in the film. Discrimination between the film and the substrate may be increased by choosing the frequency to be just above the absorption edge of the particular atom of interest. The effect of absorption of the X-rays may be detected by monitoring the fluorescence as electrons drop to lower energy levels, by collecting emitted electrons, or by measuring effects caused by electrons excited within the material. The technique has been used successfully by Nakagiri et al. (1985) to determine the distance between metal ions in a heterostructure fatty acid salt film.

5.3   Neutron reflection

The use of a neutron beam gives rise to interference effects in a similar way to an X-ray beam (Tredgold, 1994). The major difference between the two types of radiation lies in the factors governing the intensity of diffraction. For X-rays, this depends on the electron density variation across the layers. In the case of neutrons, it is the variation of nuclear scattering length density that determines the Bragg intensity. Because neutron scattering is a nuclear property, it may vary considerably from one element to the next and is different for different isotopes, e.g., hydrogen and deuterium. Neutron absorption is usually negligible and the interference effects that cause X-ray scattering to diminish with increasing angle are absent with neutrons and the scattering is isotropic.

Figure 5.10 contrasts neutron and X-ray data for fatty acid LB layers (Grundy et al., 1990). The films were built up by alternating the cadmium

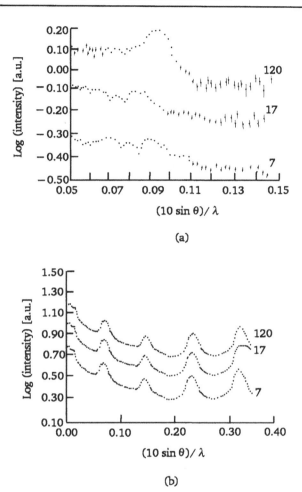

Figure 5.10 (a) Neutron and (b) X-ray reflectivity profiles for alternate-layer fatty acid salt films. The data are shown for films built-up at deposition speeds of 7, 17 and 120 mm min^{-1}. (Reprinted with permission from Grundy, M. J., Musgrove, R. J., Richardson, R. M., Roser, S. J. & Penfold, J. (1990) Langmuir, 6, 519–21. Copyright 1990 American Chemical Society.)

salts of hydrogenous and deuterated n-docosanoic acid. The X-ray reflectivity profiles shown in figure 5.10(b) are for 20-layer thick samples and the three curves correspond to films deposited at different dipping *speeds* (7, 17 and 120 mm min^{-1}). The Bragg peaks all have identical d_{001} spacings of 6 nm, consistent with the head-to-head deposition of upright molecules. However, the neutron data (figure 5.10(a)) are quite different. The curves consist of both Bragg peaks and Kiessig fringes (the measurements are made just above the

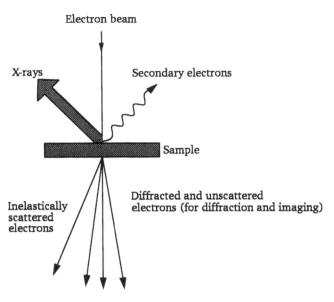

Figure 5.11 Interaction of an electron beam with a sample.

critical angle for neutrons). It is only the sample deposited at the fast speed $(120\,mm\,min^{-1})$ that has an intense 001 Bragg reflection. This suggests that intermixing of molecular layers occurs, either during the transfer of the floating monolayer to the substrate or while the substrate is under water (section 3.3).

The spacing of the Kiessig interference fringes may be used, in a similar way to X-ray effects, to obtain a measurement of the total thickness of a multilayer film (Highfield et al., 1983).

X-ray and neutron reflection measurements may also be made from floating monolayers (Barton et al., 1988; Grundy et al., 1988; Kenn et al., 1991).

5.4 Electron microscopy

Electrons are generated in an electron microscope by thermionic emission from a metal filament and accelerated through a potential. For an accelerating voltage of 100 kV, the electron wavelength is 3.7×10^{-3} nm. Atomic resolution should therefore be possible.

The main interactions taking place when the electron beam is incident on matter are shown in figure 5.11. Unscattered and diffracted electrons form the basis for conventional *transmission electron microscopy* (TEM) and diffraction. Diffracted electrons will form a pattern that can be transformed directly into an image by a magnetic lens. Either the diffraction pattern or the image can be projected onto a viewing screen.

Low energy (<50 eV) secondary electrons emitted from the surface of the sample can be used for *scanning electron microscopy*. The beam can be concentrated to a small probe (≈2 nm in diameter) that may be deflected across the sample in a raster fashion using scanning coils. Secondary electrons can be detected above the sample and an image showing the intensity of secondary electrons emitted from different parts of the sample can be built up.

A related technique involves the collection of electrons transmitted through the specimen during scanning. These can be used to produce a *scanning transmission* (STEM) image, with the advantage, compared with TEM, that radiation damage is reduced because the beam is not stationary. Emitted X-rays are characteristic of the elements in the sample and may be used for elemental analysis.

In the electron microscope, the electrons are scattered by the atomic potentials of the atoms in the sample. Generally, the scattering factor increases with atomic number. In contrast to X-rays and neutrons, the scattering of electrons by matter is very strong and diffraction is even feasible with gaseous samples.

5.4.1 Sample preparation

TEM studies of LB films require the layers either to be removed from the substrate or to be deposited onto amorphous supports (Petty, 1990). A relatively simple sample preparation technique involves the deposition of the LB structure onto a previously anodically oxidized aluminium substrate. The LB film on its alumina support is then removed by etching the aluminium layer in a mercuric chloride solution. Alternatively, LB films may be floated from glass substrates in highly diluted hydrofluoric acid and picked up on a carbon-coated copper microscope grid, or transferred to carbon-coated grids by raising the latter through the condensed floating monolayer. Specimens may also be prepared using standard replication techniques. Reflection measurements have less demanding sample preparation: LB layers can be deposited onto a variety of substrates and then rapidly studied in the electron microscope.

5.4.2 Direct imaging

Using certain LB materials, it is possible to obtain direct images under the electron microscope. A transmission electron micrograph (after image enhancement) obtained for just one layer of copper tetra-*t*-butyl phthalocyanine is given in figure 5.12 (Fryer et al., 1985). The lines of molecules are well ordered but there is general bending within a domain. There is

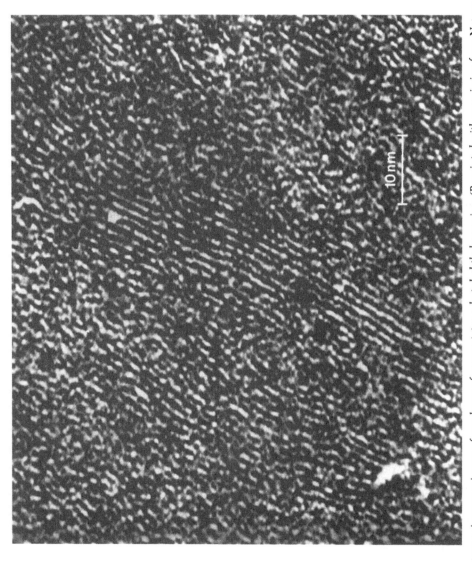

10 nm

Figure 5.12 Electron microscope image of one layer of copper tetra-t-butyl phthalocyanine. (Reprinted with permission from Nature (Fryer, J. R., Hann, R. A. & Eyres, B. L. (1985) Nature, 313, 382–4.) Copyright (1985) Macmillan Magazines Limited.)

only approximate alignment of the domains, implying that this LB film should be regarded as an amorphous matrix containing embedded crystallites.

5.4.3 Electron diffraction

Electron diffraction provides structural information on LB films in much the same way as X-ray experiments. To obtain the structure normal to the film plane, the electron beam impinges at grazing angles and the reflected beam is observed. For fatty acid type LB films, the diffraction experiments reveal the packing of the C_2H_4 subcells (section 5.1) in the aliphatic chains. The transmission geometry is used to obtain in-plane structural information. As noted above, an advantage of the electron beam compared with X-rays is that the interaction of electrons with matter is much stronger. Consequently, a diffraction pattern can be obtained with a beam of smaller diameter. Both *transmission electron diffraction* (TED) and *reflection high energy electron diffraction* (RHEED) techniques may be used (Russell, 1982). Figures 5.13(a) and (b) show TED and RHEED patterns obtained from 22-tricosenoic acid films (Peterson and Russell, 1984). Both diffraction patterns can be indexed as arising from the orthorhombic R(001) packing of the subcells (section 5.1).

The measurement of RHEED patterns in the direction of film deposition and perpendicular to it can reveal the presence of in-plane anisotropy. This is demonstrated by figure 5.14, which shows RHEED patterns obtained for a 7-layer LB film of 22-tricosenoic acid with the electron beam lying perpendicular (figure 5.14(a)) and parallel (figure 5.14(b)) to the direction of dipping (Jones et al., 1986). The rows of diffraction spots in figure 5.14(a) are tilted at approximately 20° with respect to the horizontal, showing that the molecular chains are inclined at a similar angle to the substrate normal. The sense of this tilt is upwards for samples dipped in the vertical plane. However, in figure 5.14(b), corresponding to the electron beam parallel to the dipping direction, the rows of reflections are parallel to the horizontal.

The tilt elevation for fatty acid molecules, as revealed by RHEED, is found to vary with the precise details of the LB film deposition. For example, diffraction patterns recorded from one- and three-layer films are invariant as the samples are rotated about the substrate normal, indicating no anisotropy with dipping direction for these very thin LB films (Jones et al., 1986). The patterns exhibit an unusual arrangement in which the molecular chains are all inclined at the same angle of about 20° to the substrate normal, but the direction of this tilt is random and varies from grain to grain. This is further evidence that the *molecular packing in the first few LB layers of fatty acid films can be different to that of subsequent layers* (section 3.4).

The angle of inclination to the horizontal of the line of the spots in RHEED patterns of fatty acid LB films also varies with the deposition pressure

(a)

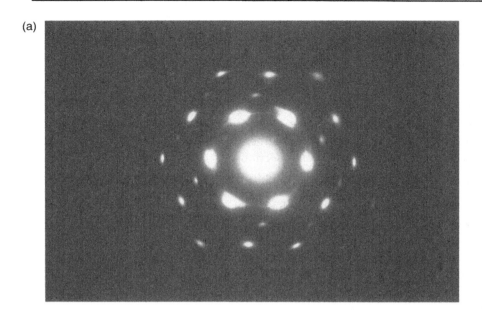

(b)

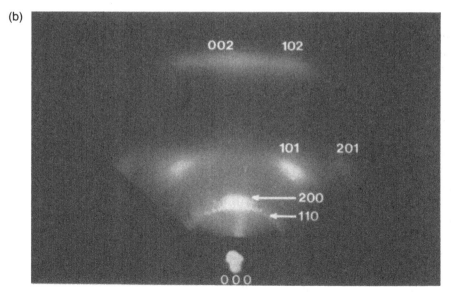

Figure 5.13 (a) Transmission electron diffraction pattern for 22-tricosenoic acid LB film. (b) Reflection high energy electron diffraction pattern for 22-tricosenoic acid. (After Peterson and Russell, 1984. Reproduced with permission from Taylor and Francis.)

(a)

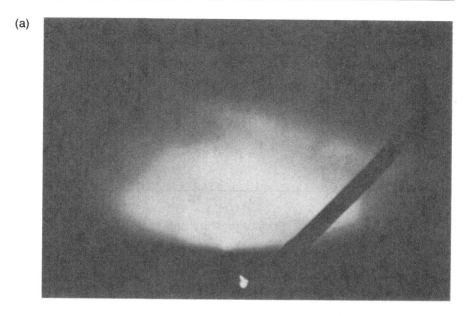

(b)

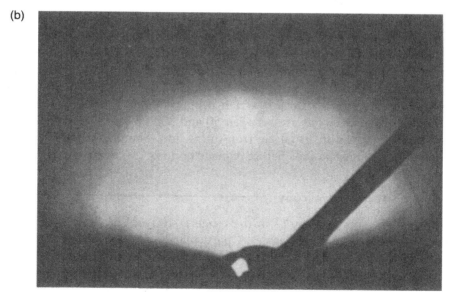

Figure 5.14 RHEED patterns from 7-layer LB films of 22-tricosenoic acid.
(a) tilted pattern obtained with the incident electron beam perpendicular to
the dipping direction. (b) pattern obtained with the electron beam parallel
to the dipping direction. (After Jones et al., 1986. Reproduced with
permission from Taylor and Francis.)

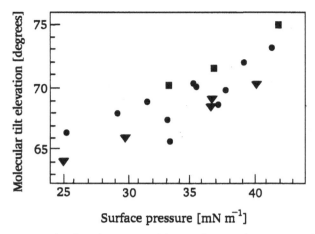

Figure 5.15 Molecular tilt (measured from substrate plane) versus deposition surface pressure for 22-tricosenoic acid. ■ deposition pH 7; ● pH 3 (After Peterson et al., 1988.) ▼ pH 7 (After Barnes and Sambles, 1987. Reproduced with permission from Elsevier Science.)

(Peterson et al., 1988; Robinson et al., 1989). Typical data are shown in figure 5.15 (Peterson et al., 1988). As the deposition pressure is increased, the molecular tilt elevation also increases (i.e., the molecules become more upright). However, this is a continuous processes and is not what would be expected from the molecules taking up one of a few fixed orientations (section 5.1).

Measurements of the inter-chain spacing in LB films may also be made using *low energy electron diffraction* (LEED) (Vogel and Wöll, 1988). The energies of the electrons are between 50 and 500 eV, much less than the kV energies employed in usual electron diffraction. Samples may therefore be exposed for longer periods before damage occurs.

5.5 Infrared spectroscopy

Infrared (IR) spectroscopy (λ in the range 1–100 µm) probes the vibrational features of an organic molecule. A particular bond must have a permanent dipole moment associated with it to interact with the infrared radiation.

The compression and extension of a bond can be likened to the behaviour of a spring and this analogy may be taken further by assuming that the bond, like a simple spring, obeys Hooke's law. For a diatomic molecule (e.g., HCl), the vibrational energy levels E_v are given by

$$E_v = h\nu(v + \tfrac{1}{2}) \tag{5.7}$$

where h is Planck's constant, ν is the frequency of the radiation and the vibrational quantum number $v = 0, 1, 2, \ldots$

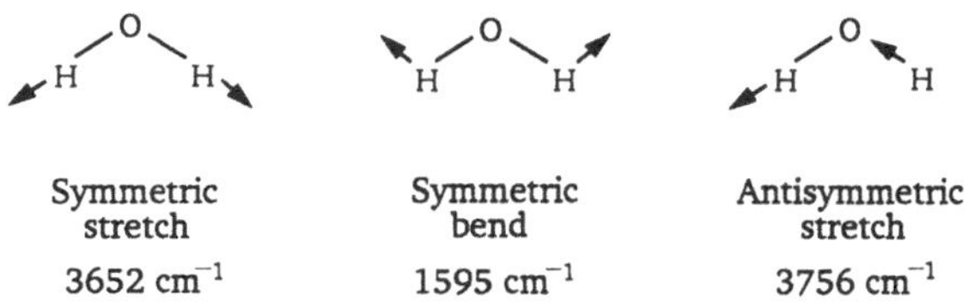

Figure 5.16 Three fundamental vibrations of the water molecule.

In IR spectroscopy, it is usual for the energy of the electromagnetic radiation to be quoted in terms of a *wavenumber* $\bar{\nu}$. This is the reciprocal of the wavelength, in centimetres, i.e.,

$$\bar{\nu} = \frac{1}{\lambda} \text{ cm}^{-1} \tag{5.8}$$

The wavenumber expresses the number of waves or cycles contained in each centimetre length of the radiation and is a useful concept in spectroscopy.

The number of vibrational modes associated with polyatomic molecules can be very large. An N-atomic molecule has $3N - 5$ *normal modes* of vibration if it is linear and $3N - 6$ if it is nonlinear (Hollas, 1992). A normal mode is one in which all the nuclei undergo harmonic motion, have the same frequency of oscillation and move in phase (but with different amplitudes). The form of these vibrations may be obtained from a knowledge of the bond lengths and angles and of the bond-stretching and angle-bending force constants. Consider, for example, a molecule of water. This is nonlinear and triatomic, with three allowed normal vibrational modes, figure 5.16. Each motion is described as *stretching* or *bending*, depending on the nature of the change in molecular shape. Furthermore, the motions are designated either *symmetric* or *antisymmetric*.

Although a normal mode of vibration involves movement of all the atoms in a molecule, there are circumstances in which movement is almost localized in one part of the molecule. If the vibration involves the stretching or bending of a terminal XY group, where X is heavy compared to Y (e.g., an OH group in a fatty acid), the corresponding vibration wavenumbers are almost independent of the rest of the molecule to which XY is attached. A typical wavenumber for the XY stretching vibration may therefore be referred to. For example, the OH stretching frequency is normally in the region of 3600 cm^{-1}. Many group vibrations occur in the region 1500 to 3700 cm^{-1}; stretching and bending vibrations of some well-known groups are listed in table 5.1.

Not all parts of a molecule are characterized by group vibrations. Many normal modes involve strong coupling between stretching or bending motions of atoms in a straight chain, a branched chain or a ring. Such vibrations are called *skeletal vibrations* and tend to be specific to a particular molecule. For this reason, the region where skeletal vibrations mostly occur,

Table 5.1 *Characteristic stretching and bending frequencies of molecular groups.*

Group	Approximate wavenumber [cm^{-1}]
$-$OH	3600
$-$NH$_2$	3400
$=$CH$_2$	3030
$-$CH$_3$	2960 (antisym. stretch)
	2870 (sym. stretch)
	1460 (antisym. bend)
	1375 (sym. bend)
$-$CH$_2-$	2920 (antisym. stretch)
	2850 (sym. stretch)
	1470 (bend)
$-$C$\equiv$C$-$	2220
$>$C$=$0	1750–1600
$>$C$=$C$<$	1650

wavenumbers below about 1400 cm^{-1}, is sometimes called the *fingerprint region*.

Besides the description of group vibrations as stretch and bend (or *deformation*), the terms *rock, twist, scissor, wag, torsion, ring breathing* and *inversion* (*umbrella*) are used frequently. Some of these are illustrated for the CH$_2$ group in figure 5.17.

Only the component of radiation polarized in the direction of the transition dipole moment will induce a change from one vibrational energy level to another. The absorption intensity of the radiation may be written

$$I \propto |\mathbf{p} \cdot \mathbf{E}|^2 \tag{5.9}$$

where **p** is the transition dipole moment (appendix B) and **E** is the electric field vector. If **E** is parallel to the transition dipole moment, the probability

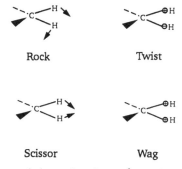

Figure 5.17 *Rocking, twisting, scissoring and wagging vibrations in a CH$_2$ group. The $\oplus$ and $\ominus$ signify movement in and out of the plane of the diagram; the arrows indicate movement in this plane.*

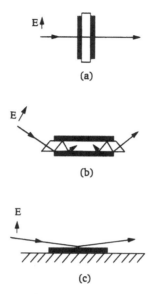

Figure 5.18 Comparison of (a) simple transmission, (b) ATR and (c) RAIRS sampling techniques. (The LB layers are shown as dark regions.)

of absorption is high, whereas if **E** is perpendicular to it no radiation is absorbed.

The absorption or reflection intensities resulting from the interaction of IR radiation with monolayer samples are very low, a direct result of the relatively small numbers of molecules being sampled. In a transmission experiment with normal incidence of the IR beam, the electric field vector is oriented parallel to the layer plane (figure 5.18(a)). In this geometry the projection of the transition moments on the layer plane is probed. An increase in surface sensitivity may be obtained by using a method based on *attenuated total reflection* (ATR), figure 5.18(b). In this technique, the LB sample is deposited onto either side of an IR-transmitting crystal (e.g., silicon or germanium). The radiation is incident at an angle greater than the critical angle (appendix B) and undergoes multiple reflections inside the crystal. On each reflection, the evanescent field of the IR beam penetrates the LB film and may be absorbed by it (the evanescent field decays over micrometre distances – section 7.3). Both the simple transmission and ATR experiments may be performed using either polarized or unpolarized radiation.

Reflection absorption infrared spectroscopy, RAIRS, requires IR radiation to be incident at a grazing angle (85°–88°) to a metal surface, onto which the LB film has been deposited, figure 5.18(c). Incident *s*-polarized radiation (appendix B) undergoes a phase shift of 180° on reflection from the metal surface and so the electric field components of the incident and reflected radiation

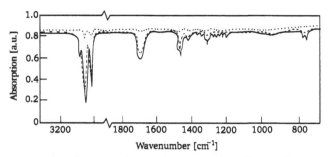

Figure 5.19 Infrared spectra of n-octadecanoic acid LB films (33 layers). ATR spectrum (p-polarized incident radiation) ———; ATR spectrum (s-polarized incident radiation) – – –; transmission spectrum ····. (After Takenaka et al., 1971. Reproduced with permission from Academic Press.)

cancel at the metal/LB film interface. In contrast, the incident and reflected components of *p*-polarized IR radiation differ by only about 90° at grazing incidence (Yarwood, 1993). This is the origin of the *surface selection rule* which results in the very useful ability to distinguish vibrations that possess a transition dipole moment with a large component perpendicular to the surface. Although unpolarized radiation is often used in RAIRS experiments, the use of *p*-polarized incident radiation can result in an improvement in the signal-to-noise ratio (Song et al., 1991). (The *s*-polarized component of unpolarized light carries no information, but contributes to the noise in the detected signal.) One disadvantage of RAIRS is that a single reflection leads to relatively poor sensitivity. However, improvements in the signal-to-noise ratio may be obtained by using a *Fourier transform* infrared (FTIR) spectrometer (Yarwood, 1991 & 1993).

Figure 5.19 compares the simple transmission and ATR spectra for LB films of n-octadecanoic acid (Takenaka et al., 1971). It is evident that the ATR technique produces a considerable improvement in signal intensity over the simple transmission measurement. The spectra reveal both the localized group modes in the CH stretching region (2800–3000 cm^{-1}) and many evenly spaced bands between 1500 and 1100 cm^{-1}. The latter are associated with CH_2 wagging and twisting and rocking. Also apparent is a strong absorption bond at 1700 cm^{-1} due to the stretching of the C=O bond. This band is not detected for multilayers of fatty acids deposited from subphases containing divalent ions, confirming that the monolayers are transferred to the substrate as the fatty acid *salt* (Rabolt et al., 1983). Close examination of the 1700 cm^{-1} band reveals that it is a doublet (1695 and 1705 cm^{-1}); the two components are related to the *cis*- and *trans*-conformations of the carboxylic acid dimers (figure 5.2). The observation of this band has also proved useful in obtaining information concerning

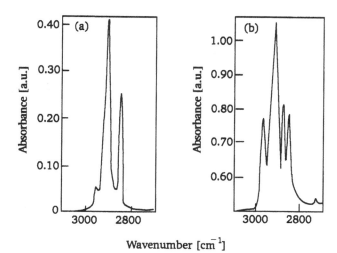

Figure 5.20 Infrared spectra of 21 LB layers of cadmium docosanoate on a silicon ATR crystal. (a) ATR mode; (b) RAIRS mode. (After Davies and Yarwood, 1989. Reproduced with permission from the American Chemical Society.)

proton transfer in fatty acid/fatty amine alternate-layer LB films (Jones et al., 1987).

One surprising fact from many infrared measurements on LB films is the absence of an absorption peak in the $3400 \, cm^{-1}$ region due to water molecules. Although, it might be expected that water is always incorporated in multilayers during the deposition process, it is rarely observed in IR experiments.

The orientation of the molecules in LB films may be investigated using different polarizations of incident radiation or by comparing ATR and RAIRS measurements. Figure 5.20 shows the ATR and RAIRS spectra (both using p-polarized radiation) in the CH stretching region for 21 layers of cadmium docosenoate (Davies and Yarwood, 1989). The bands that are observed are $2960 \, cm^{-1}$ (antisymmetric CH_3 stretch), $2920 \, cm^{-1}$ (antisymmetric CH_2 stretch), $2870 \, cm^{-1}$ (symmetric CH_3 stretch) and $2850 \, cm^{-1}$ (symmetric CH_2 stretch). The symmetric and antisymmetric CH_2 stretches are both intense in the ATR spectrum. However, in the RAIRS mode the CH_3 bands increase in intensity. The transition dipole moments of the CH_2 vibrations are perpendicular to the long axis of the fatty acid molecule, whereas the symmetric and antisymmetric CH_3 stretches have components along the chains axis. Therefore, the experimental data shown in figure 5.20 are consistent with the long axes of the fatty acid molecules being almost perpendicular to the substrate plane.

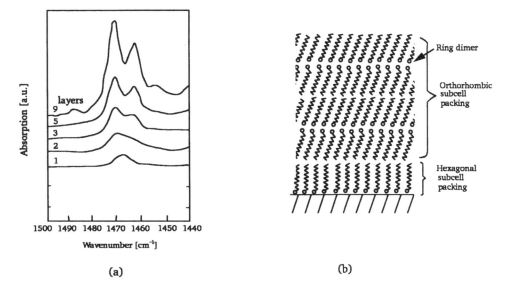

Figure 5.21 (a) FTIR-ATR spectra in the CH_2 scissoring region for different numbers of n-octadecanoic acid monolayers. (b) Proposed packing arrangements in the LB film. (Reprinted with permission from Kimura, F., Umemura, J. & Takenaka, T. (1986) Langmuir, 2, 96–101. Copyright 1986 American Chemical Society.)

Further interesting IR vibrations in long-chain fatty acid materials are those due to the CH_2 scissoring vibrations located around $1470\,cm^{-1}$. In some cases, a band appears as a singlet at $1468\,cm^{-1}$ with a symmetrical shape and in others a doublet, with maxima at 1473 and $1465\,cm^{-1}$, is evident (Chollet and Messier, 1982; Rabolt et al., 1983; Kimura et al., 1986). The band splitting is caused by a change in the crystal structure and is characteristic of an orthorhombic packing of the C_2H_4 subcells, with two molecules per unit cell. Figure 5.21(a) shows the IR region near this band for different numbers of LB layers of n-octadecanoic acid (Kimura et al., 1986). The band is a singlet for the one-monolayer film, while two vibrational bands are seen for layers containing more than three monolayers. A possible molecular arrangement to explain these data is shown in figure 5.21(b). The molecules in the first monolayer are upright and packed in a hexagonal arrangement. For thicker films, the hydrocarbon chains crystallize in dimers in the monoclinic C form (figure 5.4) with an orthorhombic subcell packing and the long axes of the molecules tilted at about 30° to the substrate normal. There is therefore some agreement with the electron diffraction results on small numbers of fatty acid layers discussed earlier (section 5.4.3).

Infrared spectroscopy may also be used to monitor chemical and structural changes occurring in multilayer films. For example, figure 5.22 shows the

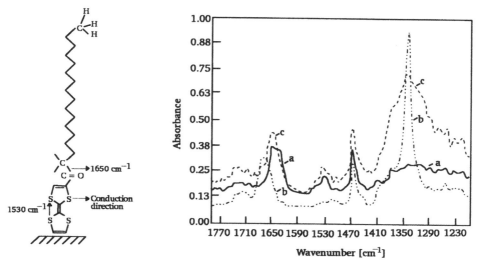

Figure 5.22 ATR spectrum of amphiphilic TTF derivative shown on the left: curve a, as-deposited; curve b, immediately after iodine doping; curve c, after 2 h in air. (Reprinted with permission from Dhindsa, A. S., Bryce, M. R., Anceln, H., Petty, M. C. & Yarwood, J. (1990) Langmuir, **6,** *1680–2. Copyright 1990 American Chemical Society.)*

results of an IR study of TTF (section 4.6) LB layers before and after doping with iodine vapour (Dhindsa et al., 1990). The molecules of this compound are known to transfer to the substrate with their long axes roughly perpendicular to the surface. The C=O stretching vibration, at $1650\,cm^{-1}$, has its transition dipole oriented perpendicular to the long axis of the molecule while the C=C stretching mode of the fulvalene ring, at approximately $1530\,cm^{-1}$, is parallel to this direction. The intensity of these bands reflects strongly the chemistry of the initial doping (oxidation) and subsequent loss of I_2 to produce an organic conductor. Immediately after doping (curve b in figure 5.22) the C=O stretching band moves to a higher frequency because of electronic changes associated with the oxidation of the TTF derivative HDTTF to produce a radical cation $HDTTF^{\cdot+}$. The C=C band ($1530\,cm^{-1}$) almost disappears, showing that the radical cation carries a largely C–C single bond. The band re-appears (curve c) when iodine molecules leave the film to form a mixed valence complex, with some fulvalene rings in their ground (not charge-transfer) state. The charge-transfer in such systems often leads to vibronic excitation of originally IR-forbidden vibrations (Hollas, 1992). Figure 5.22 shows a very intense band near $1350\,cm^{-1}$ immediately after doping with I_2. This arises from the coupling of the motion of electrons with vibrational modes in the fulvalene ring, referred to as *electron–molecule vibronic coupling.*

5.5.1 Raman scattering

When electromagnetic energy falls on an atomic or molecular sample, it may be absorbed if the energy of the radiation corresponds to the separation of two energy levels in the atoms or molecules (appendix B). If it does not, the radiation will be either transmitted or scattered. Of the scattered radiation, most is of unchanged wavelength λ and this is called *Rayleigh scattering*. The intensity I_s is related to the wavelength by

$$I_s \propto \lambda^{-4} \tag{5.10}$$

However, a small amount of the scattered radiation is at a slightly increased or decreased wavelength. This is *Raman scattering*; the radiation scattered with an energy lower than the incident beam is referred to as *Stokes' radiation*, while that at higher frequency is called *anti-Stokes' radiation*. To be Raman active, a molecular rotation or vibration must cause some change in a component of the molecular polarizability (section B.4).

Raman scattering is an inherently weak process, so an enhancement mechanism is usually needed to study LB layers (Petty, 1990). An increase in the signal can usually be achieved by depositing films onto the surfaces of noble metals such as Au or Ag (Chen et al., 1985) or by the interaction with surface plasmons (section 7.4) (Knoll et al., 1982). These methods are generally referred to as *surface-enhanced Raman spectroscopy*.

For materials with electronic transitions in the visible region, the *resonance Raman* effect can be used to enhance certain bands (Vandevyver et al., 1983; Uphaus et al., 1985). Fourier transform (FT) Raman spectroscopy can be applied to LB layers of dye molecules (Zimba et al., 1988). Since no resonance enhancement occurs when excitation is in the near IR, FT Raman allows vibrations not associated with the electronic chromophore to be observed.

The Raman process may be used as a basis for a microscopy technique. The mode of operation is similar to a fluorescence microscope. Images are recorded with a sensitive (e.g., charge-coupled device – CCD) camera in light that has been scattered in a particular Raman band using optical filters. Figure 5.23 shows a Raman image of a 4 μm by 4 μm grid produced by tracing an electron beam horizontally and vertically across a 16-layer LB film of an amphiphilic diacetylene material (Sudiwala et al., 1992). The bright areas in the photograph are regions of predominantly the blue form of the polymer (formed by proximity effects), while the darkest regions correspond to the red form, formed by direct exposure to the electron beam (section 4.7.1).

If the photon scattered by the sample in a Raman experiment is acoustic rather than optical, the process is called *Brillouin scattering*. This technique may be used to measure the elastic constants of LB films (Zanoni et al., 1985).

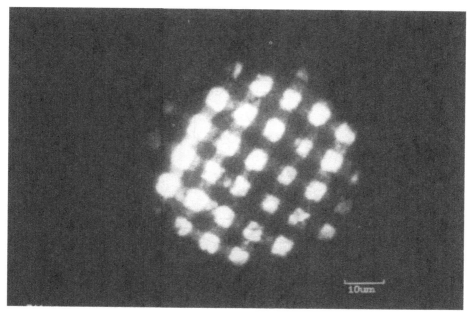

Figure 5.23 Raman image of a 4 μm by 4 μm grid formed by scanning an electron beam across a 16-layer LB film of an amphiphilic diacetylene; exposure dose $1.6 \times 10^6 \, \mu C \, m^{-2}$. (After Sudiwala et al., 1992. Reproduced with permission from Elsevier Science.)

5.6   Scanning tunnelling and atomic force microscopies

Scanning tunnelling microscopy (STM) and atomic force microscopy (AFM) are both techniques that may be used to provide direct images of metal surfaces with nanometre resolution (Heckl, 1992). STM may provide lateral and vertical resolutions of less than 0.3 nm and 0.02 nm, respectively. Furthermore, the electron energies are usually less than 3 eV, thus avoiding the film degradation which can be a problem with other imaging methods. Figure 5.24 shows the experimental set-ups for the two microscopies. Whereas STM records the overlap of the local electron density of states between tip and substrate or its modulation by adsorbate molecules in the gap, AFM measures the interatomic forces between a cantilevered spring tip and the sample surface. The image contrast in AFM is achieved by probing the elastic response of the molecules to the force exerted by the scanning tip.

Early experiments using STM on LB film samples exhibited poor reproducibility. Artifacts such as steps, domain walls or superstructures due to multiple tip effects and *Moiré patterns* made the images difficult to interpret. A major problem is that most LB films possess highly insulating regions (due

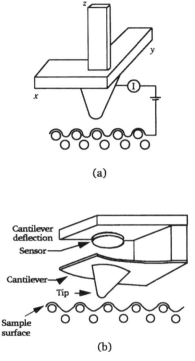

(a)

(b)

Figure 5.24 Experimental set-ups for (a) STM and (b) AFM. (After Heckl, 1992. Reproduced with permission from Elsevier Science.)

to the hydrocarbon tails) through which the tunnelling current must pass. Such difficulties are circumvented using AFM. If damage is to be avoided, care must be taken to limit the force exerted by the cantilever tip on the sample. Forces used in AFM range from 10^{-6} N in air to 10^{-11} N while imaging in liquids. Figure 5.25(a) shows an AFM image for a two-layer cadmium eicosanoate film deposited onto single crystal silicon (Chi et al., 1992). The LB film had been stored in air at room temperature for 23 days and shows grains, 70–90 nm in diameter; pinholes are also visible. The size of the grains in deposited LB films could be changed by mixing with other molecules or eliminating metal ions in the aqueous subphase. Figure 5.25(b) shows an AFM image for a one-layer eicosanoic acid film deposited on mica. The grains are much smaller than those shown in figure 5.25(a). Careful control of the AFM tip allowed patterns to be drawn in the film in figure 5.25(b). AFM has also been successfully used to provide details of the submolecular packing of the hydrophobic surface of cadmium eicosanoate LB films (Flörsheimer et al., 1994). Figure 5.26 shows an unfiltered AFM image of the surface of a 12-layer n-eicosanoic acid LB film deposited

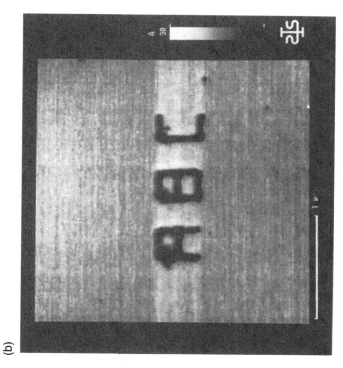

(b)

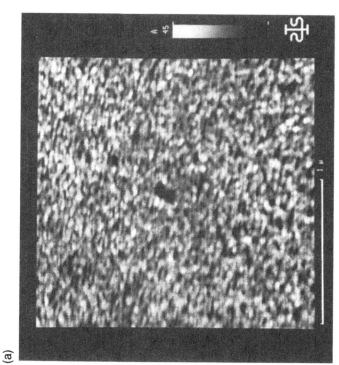

(a)

Figure 5.25 (a) AFM image of a two-layer cadmium eicosanoate LB film deposited onto Si. (b) AFM image of single LB layer of eicosanoic acid. Characters can be written into the film by controlling the AFM tip. (Reprinted with permission from Chi, L. F., Eng, L. M., Graf, K. & Fuchs, H. (1992) Langmuir, 8, 2255–61. Copyright 1992 American Chemical Society.)

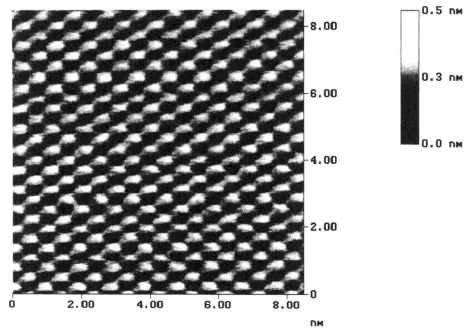

Figure 5.26 AFM image of a 12-layer n-eicosanoic acid LB film deposited onto silicon. Deposition pressure 20 mN m⁻¹; temperature 19 ± 1 °C. Unfiltered data obtained with a Digital Instruments Nanoscope III run in contact mode. (After Evenson et al., 1995.)

at a surface pressure of $20 \, \text{mN m}^{-1}$ (L'_2 phase) (Evenson et al., 1995). Lines of individual molecules are evident at the magnification shown.

5.7 Surface analytical techniques

Surface analytical techniques such as *Auger electron spectroscopy* (AES), X-ray *photoelectron spectroscopy* (XPS) and *secondary-ion mass spectrometry* (SIMS) may be used to provide useful information about the multilayer structure (Petty, 1990). These methods are contrasted in table 5.2.

Auger electron spectroscopy is a two-step process. First, a high energy electron ejects an electron from the core orbital of an atom in the material under investigation. An electron then falls down from a higher energy orbital to fill the vacancy created thereby releasing a second photoelectron, an Auger electron, from one of the higher energy orbitals. The energies of these Auger electrons are low (20–1000 eV) so that, although they may be generated from as far within the sample as the original electron beam penetrates, only those produced within the first few atomic layers of the surface can escape.

Table 5.2 *Surface analytical methods used in LB film investigations.*

Technique	Acronym	Probe beam	Sample beam	Comments
Auger electron spectroscopy	AES	high-energy (1–10 keV) electrons	electron energy (20–1000 eV)	very high surface sensitivity.
X-ray photoelectron spectroscopy	XPS	low-energy X-rays	photoelectrons	little surface damage; chemical composition can be determined.
Secondary-ion mass spectrometry	SIMS	pulsed ion (Ar$^+$) beam (few keV)	secondary ions	identification of chemical compounds.

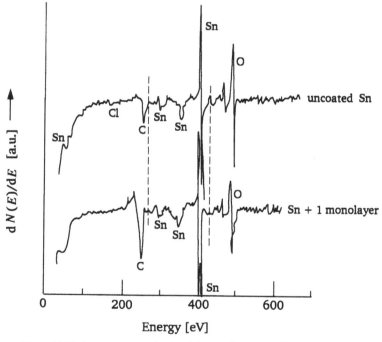

Figure 5.27 *Auger spectra obtained from a base Sn substrate and Sn + one monolayer of calcium octadecanoate, illustrating cation exchange with a reactive metal substrate. The dashed lines show where calcium peaks would be expected. (After Ginnai, 1982.)*

Therefore, the technique has immense surface sensitivity. Figure 5.27 shows an AES spectrum for a single layer of calcium octadecanoate deposited on a tin substrate and the spectrum for the substrate itself (Ginnai, 1982). The most noticeable feature is the complete absence of any structure that could be associated with calcium in the monolayer. This shows that cation exchange takes place during monolayer transfer, resulting in the substrate becoming coated with tin octadecanoate. If the calcium octadecanoate is coated onto a noble metal (e.g., gold) substrate, no ion exchange is possible and evidence for calcium is observed in the AES spectrum. Such effects were originally noted by Barraud et al. (1980) and suggest that the first monolayer in an LB array can be attached to the substrate by a strong chemical bond.

In the XPS experiment, the sample surface is irradiated by a source of low-energy X-rays. Photoionization takes place in the sample producing photo-electrons of a characteristic energy distribution. Because X-rays do not normally cause appreciable surface damage, XPS is usually preferred to AES and SIMS for organic materials. Analysis of the kinetic energy of the photoelectrons permits the elemental composition of the surface layers (≈ 10 nm) to be determined quantitatively. Information concerning the bonding environments of the elements may also be obtained.

Secondary ion mass spectrometry involves bombarding a sample surface with a pulsed primary ion beam (usually Ar^+) with an energy of a few kilo-electronvolts. This results in the emission of both positively and negatively charged secondary ions from the uppermost surface layers (a few nano-metres). The identification of chemical compounds is possible by the detection of either molecular ions with masses up to 10^4 atomic mass units or by the detection of characteristic fragments.

The relative position of a particular atom with respect to its nearest neighbours can sometimes be determined with great precision from the *extended X-ray absorption fine structure* (EXAFS). Above an X-ray absorption edge, the absorption by the atom is slightly affected by the waves being scattered by neighbouring atoms. This is manifested by weak oscillations in the absorption as a function of frequency for a considerable range above the edge. Just above the absorption edge there are stronger oscillations, called the *X-ray absorption near-edge structure* (XANES) which derive from the same process of scattering from neighbouring atoms. Both methods have been used to study the local structure and bonding states in LB films of a dye (Oyanagi et al., 1985). *Near edge X-ray absorption fine structure* (NEXAFS) is another method using synchrotron radiation that may be used to obtain orientation information about the molecules in LB layers (Rabe et al., 1988).

5.8   Film thickness measurements

Film thickness is a very important parameter in the characterization of Langmuir–Blodgett films. Many different measurement techniques have been used and some of the most popular are listed in table 5.3 (Petty, 1990). However, some of these methods do not provide an independent

Table 5.3 *Methods for measuring LB film thickness.*

Technique	Comments
X-ray diffraction	Provides d_{001} lattice spacing. Section 5.2.
Ellipsometry	Commercial instruments available; sample refractive index should be known for high accuracy. Chapter 7.
Surface plasmon resonance	Sample refractive index should be known. Chapter 7.
Capacitance versus number of layers	Provides dielectric thickness. Requires insulating samples. Possibility of damage to organic layer during metallization. Chapter 6.
Mechanical probe	Commercial instruments available. Provides metric thickness directly. Organic film may be damaged. Not suitable for monolayer sensitivity.

measure of film thickness; other physical parameters must be first determined. For example, some optical techniques (ellipsometry and surface plasmon resonance are discussed in chapter 7) measure the optical thickness (or optical path length), which is equal to the metric thickness multiplied by the refractive index. Electrical measurements based on measuring the capacitance of a metal/LB film/metal structure as a function of the number of monolayers (chapter 3, figure 3.4; section 6.6.1) yield the *dielectric thickness* (metric thickness ÷ relative permittivity). Direct measurement of the thickness of an LB film is conveniently accomplished using a mechanical probe; however, care must be taken to avoid damage to the relatively soft organic layer.

5.9 Other methods

Many analytical methods noted in this chapter require quite sophisticated (and expensive) instrumentation. However, simple approaches may be used to gain an insight into the general quality of an LB film. A rapid assessment may be made by observing the organic layer between crossed polarizers in an optical microscope. For reasonable sensitivity this technique requires film thicknesses of roughly 500 nm. In such films, essentially just one packing and one orientation of the crystallites can be resolved. Birefringent (appendix B) behaviour in fatty acids is ascribed to the presence of one of the tilted structures. The precise orientation is partly determined by the initial monolayer, which fixes the set of packings and orientations available for subsequent epitaxial enhancement. Figure 3.6 in chapter 3 is an example of this. The limitations on resolution and depth of field inherent in conventional microscopy can be improved by scanning (Hudson et al., 1992).

There are comparatively few reports of the mechanical properties of monolayer and multilayer films, although clearly these have important implications in research areas such as lubrication and permeability (Petty, 1990).

Other analytical techniques that may be used to probe the structure of multilayer LB films include: *electron spin resonance* (ESR); *electron nuclear double resonance* (ENDOR); *photoacoustic spectroscopy*; and *differential scanning calorimetry* (Petty, 1990; Ulman, 1991).

References

Barnes, W. L. and Sambles, J. R. (1987) Surface pressure effects on Langmuir–Blodgett multilayers of 22-tricosenoic acid, *Surf. Sci.*, **187**, 144–52

Barraud, A., Rosilio, C. and Ruaudel-Teixier, A. (1980) Reactivity of organic molecules in monolayers, *Thin Solid Films*, **68**, 7–12

Barton, S. W., Thomas, B. N., Flom, E. B., Rice, S. A., Lin, B., Peng, J. B., Ketterson, J. B. and Dutta, P. (1988) X-ray diffraction study of a Langmuir monolayer of $C_{21}H_{43}OH$, *J. Chem. Phys.*, **89**, 2257–70

Belbeoch, B., Roulliay, M. and Tournarie, M. (1985) Evidence for chain interdigitation in Langmuir–Blodgett films, *Thin Solid Films*, **134**, 89–99

Cheetham, A. K. (1988) Diffraction methods, in *Solid State Chemistry Techniques*, eds. A. K. Cheetham and P. Day, Oxford Sci. Pub., Oxford

Chen, Y. J., Carter, G. M. and Tripathy, S. K. (1985) Study of Langmuir–Blodgett polydiacetylene polymer films by surface enhanced Raman scattering, *Solid State Commun.*, **54**, 19–22

Chi, L. F., Eng, L. M., Graf, K. and Fuchs, H. (1992) Structure and stability of Langmuir–Blodgett films investigated by scanning force microscopy, *Langmuir*, **8**, 2255–61

Chollet, P. A. and Messier, J. (1982) Studies of oriented Langmuir–Blodgett multilayers by infrared linear dichroism, *Chem. Phys.*, **73**, 235–42

Davies, G. H. and Yarwood, J. (1989) Infrared intensity enhancement for Langmuir–Blodgett monolayers using thick metal overlayers, *Langmuir*, **5**, 229–32

Dhindsa, A. S., Bryce, M. R., Ancelin, H., Petty, M. C. and Yarwood, J. (1990) Infrared spectroscopic studies on the structure and ordering of hexadecanolytetrathiafulvalene conducting Langmuir–Blodgett multilayers, *Langmuir*, **6**, 1680–2

Evenson, S. A., Pearson, C., Badyal, J. P. and Petty, M. C. (1995) unpublished data

Feigin, L. A., Lvov, Y. M. and Troitsky, V. I. (1989) X-ray and electron diffraction study of Langmuir–Blodgett films, *Sov. Sci. Rev. A Phys.*, **11**, 285–377

Feynman, R. P., Leighton, R. B. and Sands, M. (1963) *The Feynman Lectures on Physics*, Vol. 1, Addison-Wesley, Reading, MA

Flörsheimer, M., Steinfort, A. J. and Günter, P. (1994) Submolecular details of Cd arachidate Langmuir–Blodgett films detected by atomic force microscopy, *Thin Solid Films*, **244**, 1078–82

Fryer, J. R., Hann, R. A. and Eyres, B. L. (1985) Single organic monolayer imaging by electron microscopy, *Nature*, **313**, 382–4

Ginnai, T. M. (1982) An investigation of electron tunnelling and conduction in Langmuir films, PhD thesis, Leicester Polytechnic, UK

Grundy, M. J., Richardson, R. M., Roser, S. J., Penfold, J. and Ward, R. C. (1988) X-ray and neutron reflectivity from spread monolayers, *Thin Solid Films*, **159**, 43–52

Grundy, M. J., Musgrove, R. J., Richardson, R. M., Roser, S. J. and Penfold, J. (1990) Effect of dipping rate on alternating layer Langmuir–Blodgett film structure, *Langmuir*, **6**, 519–21

Heckl, W. M. (1992) Scanning tunnelling microscopy and atomic force microscopy on organic and biomolecules, *Thin Solid Films*, **210/211**, 640–7

Highfield, R. R., Thomas, R. K., Cummins, P. G., Gregory, D. P., Mingins, J., Hayter, J. B. and Schärpf, O. (1983) Critical reflection of neutrons from Langmuir–Blodgett films on glass, *Thin Solid Films*, **99**, 165–72

Hollas, J. M. (1992) *Modern Spectroscopy*, 2nd edition, Wiley, Chichester

Hudson, A. J., Wilson, T. and Roberts, G. G. (1992) Scanning optical microscopy of Langmuir–Blodgett films, *Thin Solid Films*, **210/211**, 689–92

Jones, C. A., Russell, G. J., Petty, M. C. and Roberts, G. G. (1986) A reflection high-energy electron diffraction study of ultra-thin Langmuir–Blodgett films of 22-tricosenoic acid, *Philos. Mag. B*, **54**, L89–L94

Jones, C. A., Petty, M. C., Roberts, G. G., Davies, G., Yarwood, J., Ratcliffe, N. M.. and Barton, J. W. (1987) IR studies of pyroelectric Langmuir–Blodgett films, *Thin Solid Films*, 155, 187–95

Kapp, D. S. and Wainfan, N. (1965) X-ray interference structure in the scattered radiation from barium stearate multilayer films, *Phys. Rev.*, 138, A1490–A1495

Kenn, R. M., Böhm, C., Bibo, A. M., Peterson, I. R., Möhwald, H., Als-Nielsen, J. and Kjaer, K. (1991) Mesophases and crystalline phases in fatty acid monolayers, *J. Phys. Chem.*, 95, 2092–7

Kimura, F., Umemura, J. and Takenaka, T. (1986) FTIR-ATR studies on Langmuir–Blodgett films of stearic acid with 1–9 monolayers, *Langmuir*, 2, 96–101

Kitaigorodskii, A. I. (1961) *Organic Chemical Crystallography*, Consultants Bureau, New York

Knoll, W., Philpott, M. R., Swalen, J. D. and Girlando, A. (1982) Surface plasmon enhanced Raman spectra of monolayer assemblies, *J. Chem. Phys.*, 77, 2254–60

Leuthe, A. and Riegler, H. (1992) Thermal behaviour of Langmuir–Blodgett films. II: x-ray and polarized reflection microscopy studies on coexisting polymorphism, thermal annealing and epitaxial layer growth of behenic acid multilayers, *J. Phys. D: Appl. Phys.*, 25, 1786–97

Lvov, Y., Svergun, D., Feigin, L. A., Pearson, C. and Petty, M. C. (1989) Small angle x-ray analysis of alternate-layer Langmuir–Blodgett films, *Philos. Mag. Lett.*, 59, 317–23

Malta, V., Celotti, G., Zannetti, R. and Martelli, A. F. (1971) Crystal structure of the C form of stearic acid, *J. Chem. Soc. B*, 13, 548–53

Mann, B. and Kuhn, H. (1971) Tunnelling through fatty acid salt monolayers, *J. Appl. Phys.*, 42, 4398–405

Matsuda, A., Sugi, M., Fukui, T., Iizima, S., Miyahara, M. and Otsubo, Y. (1977) Structure study of multilayer assembly films, *J. Appl. Phys.*, 48, 771–4

Nakagiri, T., Sakai, K., Iida, A., Ishikawa, T. and Matsushita, T. (1985) X-ray standing wave method applied to the structural study of Langmuir–Blodgett films, *Thin Solid Films*, 133, 219–55

Nicklow, R. M., Pomerantz, M. and Segmüller, A. (1981) Neutron diffraction from small numbers of Langmuir–Blodgett monolayers of manganese stearate, *Phys. Rev. B*, 23, 1081–7

Oyanagi, H., Sugi, M., Kuroda, S-I., Iizima, S., Ishiguro, T. and Matsushita, T. (1985) Polarized X-ray absorption spectra of Langmuir–Blodgett films: local structure studies of merocyanine dyes, *Thin Solid Films*, 133, 181–8

Peterson, I. R. and Russell, G. J. (1984) An electron diffraction study of ω-tricosenoic acid Langmuir–Blodgett films, *Philos. Mag. A*, 49, 463–73

Peterson, I. R. and Russell, G. J. (1985a) The deposition and structure of Langmuir–Blodgett films of long-chain acids, *Thin Solid Films*, 134, 143–52

Peterson, I. R. and Russell, G. J. (1985b) Deposition mechanisms in Langmuir–Blodgett films, *Br. Poly. J.*, 17, 364–7

Peterson, I. R., Russell, G. J., Earls, J. D. and Girling, I. R. (1988) Surface pressure dependence of molecular tilt in Langmuir–Blodgett films of 22-tricosenoic acid, *Thin Solid Films*, 161, 325–31

Petty, M. C. (1990) Characterization and properties, in *Langmuir–Blodgett Films*, ed. G. G. Roberts, pp. 133–221, Plenum Press, New York

Pomerantz, M. and Segmüller, A. (1980) High resolution X-ray diffraction from small numbers of Langmuir–Blodgett layers of manganese stearate, *Thin Solid Films*, 68, 33–45

Prakash, M., Ketterson, J. B. and Dutta, P. (1985) Study of in-plane structure in lead-fatty acid LB films using X-ray diffraction, *Thin Solid Films*, 134, 1–4

Rabe, J. P., Swalen, J. D. and Outka, D. A. (1988) Near-edge fine structure studies of oriented molecular chains in polyethylene and Langmuir–Blodgett monolayers on Si(111), *Thin Solid Films*, 159, 275–83

Rabolt, J. F., Burns, F. C., Schlotter, N. E. and Swalen, J. D. (1983) Anisotropic orientation in molecular monolayers by infrared spectroscopy, *J. Chem. Phys.*, 78, 946–52

Robinson, I., Sambles, J. R. and Peterson, I. R. (1989) A reflection high energy electron diffraction analysis of the orientation of the monoclinic subcell of 22-tricosenoic acid Langmuir–Blodgett bilayers as a function of the deposition pressure, *Thin Solid Films*, 172, 149–58

Russell, G. J. (1982) Practical reflection electron diffraction, *Prog. Crystal Growth and Charact.*, 5, 291–321

Russell, G. J., Petty, M. C., Peterson, I. R., Roberts, G. G., Lloyd, J. P. and Kan, K. K. (1984) A RHEED study of cadmium stearate Langmuir–Blodgett films, *J. Mat. Sci. Letts.*, 3, 25–8

Song, Y. P., Petty, M. C. and Yarwood, J. (1991) Effects of polarization of infrared spectra collected in reflection at grazing incidence, *Vibrational Spectroscopy*, 1, 305–9

Srivastava, V. K. and Verma, A. R. (1966) Interferometric and X-ray diffraction study of 'built-up' molecular films of some long-chain compounds, *Solid State Commun.*, **4**, 367–71

Sudiwala, R. V., Cheng, C., Wilson, E. G. and Batchelder, D. N. (1992) Polydiacetylene mono- and multilayer Langmuir–Blodgett films as electron beam resists: scanning electron microscope and Raman images, *Thin Solid Films*, **210/211**, 452–4

Takenaka, T., Nogami, K., Gotoh, H. and Gotoh, R. (1971) Studies on built-up films by means of the polarized infrared ATR spectrum. I. Built-up films of stearic acid, *J. Colloid Interface Sci.*, **35**, 395–402

Tippmann-Krayer, P., Möhwald, H. and Lvov, Y. M. (1991) Structural changes before and during desorption of Langmuir–Blodgett films, *Langmuir*, **7**, 2298–302

Tredgold, R. H. (1994) *Order in Thin Organic Films*, Cambridge University Press, Cambridge

Ulman, A. (1991) *Ultrathin Organic Films*, Academic Press, San Diego

Ungar, G. (1983) Structure of rotator phases in n-alkanes, *J. Phys. Chem.*, **87**, 689–95

Uphaus, R. A., Cotton, T. M. and Möbius, D. (1985) Surface-enhanced resonance Raman spectroscopy of synthetic dyes and photosynthetic pigments in monolayer and multilayer assemblies, *Thin Solid Films*, **132**, 173–84

Vandevyver, M., Ruaudel-Teixier, A., Brehamet, L. and Lutz, M. (1983) Polarized resonance Raman spectroscopy of Langmuir–Blodgett films, *Thin Solid Films*, **99**, 41–4

Vogel, V. and Wöll, C. (1988) Structure and dynamics of deposited lipid monolayers: low energy electron diffraction and scattering of thermal energy helium atoms, *Thin Solid Films*, **159**, 429–34

Yarwood, J. (1991) Infrared spectroscopy of Langmuir–Blodgett multilayers, *Colloids and Surfaces*, **52**, 35–46

Yarwood, J. (1993) Fourier transform infrared reflection spectroscopy for surface analysis, *Analytical Proceed.*, **30**, 13–18

Zanoni, R., Naselli, C., Bell, G., Stegeman, R., Sprague, R., Seaton, C. and Lindsay, S. (1985) Brillouin spectroscopy of Langmuir–Blodgett films, *Thin Solid Films*, **134**, 179–86

Zimba, C. G., Hallmark, V. M., Rabolt, J. F. and Swalen, J. D. (1988) Fourier transform Raman spectroscopy of thin films, *Thin Solid Films*, **160**, 311–16

6

Electrical phenomena

6.1 Basic ideas

When a d.c. *electric field* E is applied to a material, a *current* will flow. If *Ohm's law* is obeyed the *current density* **J** (current per unit cross-section of material) will be proportional to the applied field. The constant of proportionality is the *conductivity* σ

$$\mathbf{J} = \sigma\mathbf{E} \tag{6.1}$$

The *mobility* μ determines the (additional) velocity that a charge carrier (an electron or an ion) acquires because of the electric field. It is defined as the *drift velocity* $\mathbf{v_d}$ per unit field, i.e.,

$$\mu = \frac{\mathbf{v_d}}{\mathbf{E}} \tag{6.2}$$

The mobility is further related to the conductivity by the expression

$$\sigma = |q|n\mu \tag{6.3}$$

where n is the *density of charge carriers* and $|q|$ is the magnitude of their charge (charge on an electron $= 1.6 \times 10^{-19}$ C).

Organic materials with completely saturated chemical bonds (section 4.1.2) possess few charge carriers to respond to an applied electric field. Therefore, substances such as polyethylene are highly insulating, with conductivity values of about $10^{-18}\,\Omega^{-1}\,m^{-1}$ (or $S\,m^{-1}$) at room temperature (cf. copper, $\sigma \approx 10^{8}\,\Omega^{-1}\,m^{-1}$). The carrier mobilities in such compounds can also be quite small ($<10^{-10}\,m^{2}\,V^{-1}\,s^{-1}$) (Blythe, 1979). Although the application of an electric field may give rise to a very small direct current, the most important effect in these highly insulating, or *dielectric*, materials will be a displacement of fixed charges and the creation of dipoles – the process of *polarization*.

131

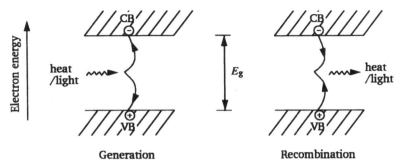

Figure 6.1 Band model of a semiconductor with a forbidden energy gap E_g
*showing the generation and recombination of charge carriers. (CB —
conduction band; VB — valence band.)*

If the molecules contain delocalized π-electrons (appendix A), conductivities similar to those of inorganic semiconductors ($\approx 10^{-2}\,\Omega^{-1}\,m^{-1}$) may be observed. Metallic and superconducting behaviour can even be achieved in certain compounds (Ferraro and Williams, 1987). The electronic energy levels in these materials are conveniently described using band theory (appendix A) in which carriers are confined to bands of energy separated by a forbidden energy gap, as shown in figure 6.1. The electrical conduction results from the movement of electrons in the conduction band and holes (appendix A) in the valence band. At low temperatures, there are few carriers available. However as the temperature is increased, electrons and holes are generated by thermal energy and the conductivity increases. Electrons and holes will also recombine and a dynamic equilibrium between generation and recombination is established at any temperature. There is an exponential relationship between the number of charge carriers $n(T)$ and temperature T

$$n(T) = n(0)\exp(-\Delta E/kT) \qquad (6.4)$$

where $n(0)$ is a constant, ΔE is an activation energy, related to the magnitude of the forbidden energy gap and k is Boltzmann's constant ($1.381 \times 10^{-23}\,J\,K^{-1}$). If the carrier mobility changes little with temperature (compared to the exponential variation in n), the electrical conductivity variation with temperature may be approximated as exponential (Okamoto and Brenner, 1964).

The number of charge carriers can also be increased by irradiating the sample with electromagnetic radiation of a suitable wavelength. If the energy of the photons $h\nu$ is less than that of the band gap E_g then no inter-action takes place and the light simply passes through the sample, which appears transparent at that wavelength. However, if the photon energy is equal to, or greater than, E_g, the light will be absorbed, with a consequent generation of electrons and holes. This is the origin of *photoconductivity* in semiconducting materials.

The electrical processes discussed above are referred to as *intrinsic* conductivity since they are related to the properties of the pure material. When impurities are added to a semiconductor, by doping (appendix A), these may dominate the electrical behaviour and the conductivity is then termed *extrinsic*.

In equation (6.1), **J** and **E** are vectors that can each be resolved into three components (J_1, J_2, J_3 and E_1, E_2, E_3) along mutually orthogonal axes. This makes σ a tensor quantity, similar to permittivity (appendix B). For a completely isotropic material only one conductivity value will be needed to characterize the material. However, for an anisotropic material, such as an LB multilayer array, *the conductivity will depend on the direction in which the electric field is applied.*

Figure 6.2 shows two different configurations that are often used for LB film studies. The electrical conductivity can either be measured parallel to the substrate surface (i.e., in the film plane) or perpendicular to it (through-plane). In the latter case the conductivity is usually dominated by the highly insulating hydrocarbon chains. Pin-holes and other defects in the organic multilayer markedly affect the conductivity values that are measured in this direction.

Direct current conduction and high frequency phenomena are essentially separate and, for the most part, independent quantities. However, at low frequencies, both can contribute to a measured conductance. The frequency-dependent conductivity $\sigma(\omega)$ of a sample may be expressed in the form

$$\sigma(\omega) = \sigma_{dc} + \sigma_{ac}(\omega) \tag{6.5}$$

where σ_{dc} is the d.c. conductivity discussed above and $\sigma_{ac}(\omega)$ is the a.c. component.

The a.c. conductivity of a sample is generally measured by applying a voltage at an appropriate frequency and then measuring both the in-phase and 90° out-of-phase components of current. The equipment required depends on the frequency of measurement: simple bridge circuits can be used over the frequency range 1 to 10^7 Hz; for lower frequencies, step-response methods are used (Harrop, 1972). The out-of-phase component of current is proportional to the real part of the sample's relative permittivity, ϵ'_r; this provides a measurement of capacitance (see section 6.6.1). The in-phase conductivity is related to the imaginary part ϵ''_r by

$$\sigma_{ac}(\omega) = \omega\epsilon_0\epsilon''_r(\omega) \tag{6.6}$$

where ϵ_0 is the permittivity of free space (8.854×10^{-12} F m^{-1}). If the sample is lossless over the frequency range of measurement, i.e., it exhibits no *relaxations*

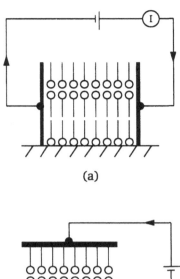

(a)

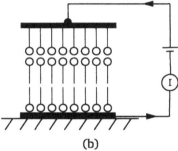

(b)

Figure 6.2 Techniques for measuring electrical conductivity in LB film structures: (a) in-plane; (b) through-plane.

or *resonances*, then ϵ_r'' will be constant and the a.c. conductivity will increase linearly with frequency (Harrop, 1972). However, many inorganic and organic samples exhibit a simple power-law relationship of the form

$$\sigma_{ac}(\omega) \propto \omega^n \tag{6.7}$$

where n is less than unity. This agrees with the 'Universal' model for the response of dielectric materials (Jonscher, 1983).

6.2 Practical considerations

Electrical measurements on LB layers (particularly monolayers) are a stringent test of film quality and possibly the most difficult of all the available characterization tools. Although there have been many reports on the structural and optical properties of LB layers over the last fifty years, it is only comparatively recently that workers have started to obtain reproducible and reliable electrical data; this is almost certainly due to improved film preparation and handling techniques.

An important question that any experimentalist undertaking electrical investigations should ask is: *do the conductivity values obtained accurately reflect the electrical properties of the organic molecules in the LB array or do defects, electrodes, the substrate or the surroundings influence the data?* One advantage of the LB technique is that the film thickness can easily be varied (even on the same substrate). Therefore, the effect of substrate and metal electrodes can be explored. Some of the problems most frequently encountered are discussed in the following sections.

6.2.1 Electrodes and measurement procedure

Nearly all experiments to investigate the electrical properties of LB layers require the films to be in direct contact with two solid electrodes. Typical arrangements for both through-plane and in-plane measurements are shown in figure 6.3. The structure in figure 6.3(a) facilitates easy contacting to both the top and bottom electrodes. It is fabricated by first thermally evaporating a metal, such as aluminium, onto part of a glass substrate. The LB film is then coated over this. Finally, the top contacts are established by evaporating another metal layer through a contact mask to provide the 'lollipop' shaped electrodes. This process must be carried out very carefully to avoid damaging the organic film. It is desirable to use a metal that can be deposited at a low temperature (Al, Ag, Bi or Mg are suitable) and to evaporate this layer slowly. Electrical contacts to the top and bottom electrodes may be made by soldering to the metal pads or by using an air-drying silver paste. The electrical conductivity is calculated from the values of the applied voltage V and current I using

$$\sigma = \frac{It}{VA} = \frac{t}{AR} \tag{6.8}$$

where t is the film thickness, A is the area of electrode overlap and R is the resistance. As LB layers are very thin, modest voltages can give rise to very high electric fields. For example, *100 mV applied across a 10 nm layer gives a field of $10^7\,V\,m^{-1}$*. Although this is below the electrical breakdown strength of many solid dielectrics (Harrop, 1972; Blythe 1979), breakdown may occur in LB layers at these fields because of an increase in field strength at the electrode edges or because of defects in the film.

To measure the in-plane conductivity, two metal electrodes are deposited onto an insulating (e.g., quartz) substrate and the LB film coated on top. The electrodes may also be deposited on top of the multilayer film. Figure 6.3(b) shows a system of *interdigitated electrodes*, which improves the sensitivity of this measurement. The electrical conductivity can be evaluated from the current and voltage measurements using

$$\sigma = \frac{Id}{Vlt} \tag{6.9}$$

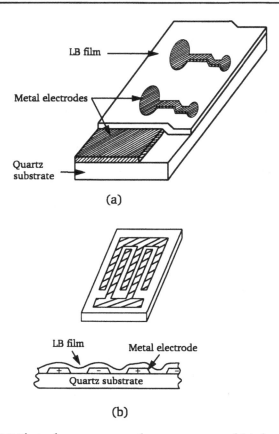

Figure 6.3 Electrode arrangements for measurement of (a) through-plane conductivity and (b) in-plane conductivity.

where d is the distance between the electrodes, l is the total length of their overlap and t is the film thickness.

In-plane measurements must be interpreted cautiously. For amphiphilic compounds incorporating long-hydrocarbon chains, the LB structure will consist of relatively conducting polar regions separated by insulating hydro-carbon planes. A metallic electrode deposited underneath or on top of the LB film will not necessarily make good electrical contact to the polar regions (a test may be made by measuring the conductivity as a function of the number of LB layers). A solution is to use a conducting silver paste for the contacts. The solvent in the paste allows the silver particles to permeate the multilayer structure. However, as such electrodes are generally established by hand (e.g., using a fine paint brush), the dimensions of the electrodes are not well defined and the electrical conductivity values will only be approximate.

Another problem for in-plane electrical characterization concerns the substrate. Clearly, when an electric field is applied in the plane of the LB film, a component of the current will flow through the substrate itself (the electrical resistance associated with the substrate will be in parallel with that due to the organic layer). Therefore, it is essential to use only high resistance substrates and to make a reference conductivity measurement using the substrate alone (with no LB film).

The electrical conductivity of most LB films is affected by the presence of moisture (Petty, 1990). This is unsurprising as the polar nature of water molecules will attract them to the head group regions of the multilayer structure. The effects can be minimized by storing the samples in a dry environment and ensuring that measurements are undertaken in a dry gas or under vacuum.

6.2.2 Equivalent circuit

The electrical *equivalent circuits* of the metal/LB film/metal structures shown in figure 6.3 are complex and will include components due to contact resistance and to any surface (oxide) layer(s) on the electrodes (these could be of the same order of thickness as the LB film!). Figure 6.4 compares such an equivalent circuit (figure 6.4(a)) with that which is actually measured (figure 6.4(b)). If d.c. (or low frequency) measurements are being made then the capacitances may be neglected. In the case of in-plane electrical measurements, the effects of contact resistance may be eliminated by using *four-contact* (or four-terminal) measurements in which a current is applied via the two outside contacts and the voltage measured using the inner electrodes (Blythe, 1979). Alternatively, a series of measurements may be made with two electrodes at different distances apart; if contact resistance effects are negligible, then the measured current, for a fixed applied voltage, should decrease linearly with the electrode separation.

Figure 6.5 illustrates how contact resistance effects can dominate the high frequency measurements of a metal/LB film/metal sandwich structure. The data show conductance values for seven monolayers of lead stearate between Al and Hg electrodes (Honig and de Koning, 1978). A careful examination of the curve reveals that the slope of the straight line varies from slightly less than unity to approximately two at a frequency close to 20 Hz. The low frequency $\sigma \propto \omega$ behaviour agrees with that expected from a dielectric material (section 6.1). However, at frequencies above 20 Hz, the measured conductance is dominated by the metallic contact resistances. A simple analysis of the equivalent circuit shows that this will result in the quadratic behaviour observed (Petty, 1990).

The nature of the equivalent circuits of electrode/LB film/electrode structures can be explored by measuring the complex *impedance* of the network

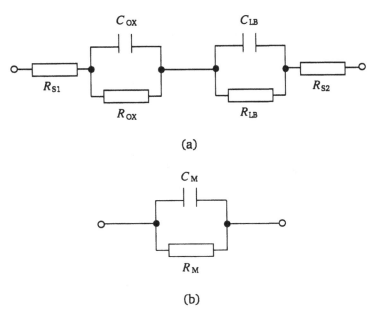

Figure 6.4 (a) Equivalent circuit of metal/LB film/metal structure. (b) Circuit actually measured. R_{S1}, R_{S2} are the contact resistances; R_{OX}, C_{OX} take into account the resistance and capacitance of oxide layer(s) on the metal electrodes; R_{LB}, C_{LB} represent the resistance and capacitance of the LB film; and R_M, C_M are the quantities measured.

over a wide frequency range (Jonscher, 1983). The technique is sometimes called *impedance spectroscopy*. The impedance Z of the simple parallel resistor–capacitor network shown in figure 6.4(b) is given by

$$\frac{1}{Z} = \frac{1}{R_M} + j\omega C_M \tag{6.10}$$

$$Z = \frac{R_M}{1 + \omega^2 C_M^2 R_M^2} - j\frac{\omega C_M R_M}{1 + \omega^2 C_M^2 R_M^2} \tag{6.11}$$

If the real part of equation (6.11) is plotted as a function of the imaginary part, over a wide frequency range, then a semicircular plot results. The addition of other series or parallel resistance–capacitance combinations results in a modification to the semicircle (e.g., addition of spurs or further semicircles).

6.2.3   Ions or electrons?

As LB layers are prepared from aqueous subphases, a contribution to the measured conductivity from ions is to be expected. It is crucial to be able to separate these ionic processes from electronic conduction. Ions travelling

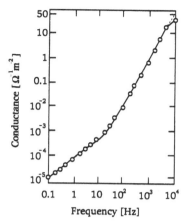

Figure 6.5 Conductance as a function of frequency for seven layers of lead stearate between an Al and an Hg electrode. (After Honig and de Koning, 1978. Reproduced with permission from the Institute of Physics.)

through a sample under the influence of an applied electric field will accumulate at defects (e.g., grain boundaries) or at one of the solid electrodes. The resulting polarization will reduce the ionic current to zero over a period of time. Ionic conductivity is expected therefore to give rise to time-dependent currents. Furthermore, ions do not respond as readily as electrons to high frequency fields and the ionic contribution to conductivity should be reduced at high frequencies. For LB materials based on charge-transfer complexes (section 4.6), the presence of a charge-transfer band in the infrared part of the electromagnetic spectrum can be used to identify electronic conductivity.

6.3 Quantum mechanical tunnelling

If the energy of an electron is less than the interfacial potential barrier in a metal/insulator/metal (MIM) structure upon which it is incident, classical physics predicts reflection of the electron at the interface. The electron cannot penetrate the barrier and its passage from one electrode to the other is inhibited. Quantum mechanics contradicts this view. The wave nature of the electron allows penetration of the forbidden region of the barrier. The *wavefunction* associated with the electron decays rapidly with depth of penetration from the electrode/insulator interface and, for barriers of macroscopic thickness, is essentially zero at the opposite interface (figure 6.6(a)), indicating a zero probability of finding an electron. However, if the barrier is very thin (<5 nm), the wavefunction has a nonzero value at the opposite interface. For this case, there is a finite probability that the electron can pass from one electrode to the other by penetrating the barrier, figure

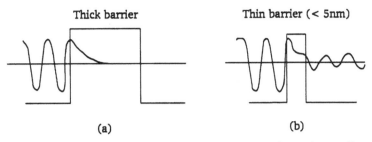

Figure 6.6 Schematic representation of quantum mechanical tunnelling: (a) thick barrier; (b) thin barrier.

6.6(b). This process is called *tunnelling* (or *quantum mechanical tunnelling*). The current versus voltage relationships for tunnelling are quite complex and depend on the magnitude of the applied voltage and whether the tunnel barrier is symmetric or asymmetric (i.e., whether the two electrodes are similar or different metals) (Simmons, 1971). For very low applied voltages (much less than the energy barrier height divided by the electronic charge), the tunnelling probability varies exponentially with the barrier thickness and the tunnelling conductivity σ_t may be given by an equation of the form

$$\sigma_t = A \exp(-Bt) \qquad (6.12)$$

where t is the thickness of the tunnel barrier and A, B are constants.

LB films of simple fatty acids have a well-defined thickness and should be an excellent basis for studying quantum mechanical tunnelling. With careful attention to experimental detail, it is possible to observe some of the theoretical predictions for tunnelling. Figure 6.7 shows the results of experiments of Polymeropoulos (1977). The tunnelling conductivity (measured at an applied voltage of 10 mV) dependence upon film thickness is clearly of the form of equation (6.12). For larger applied voltages, the current versus voltage behaviour for fatty acid monolayer films is also similar to that predicted by theory (Petty, 1990).

Some interpretations of tunnelling processes in LB films can be criticized because not enough importance has been attached to the role of the interfacial (oxide) layers associated with the metal electrodes. In view of the large quantity of experimental data, showing that the tunnelling conductivity depends strongly upon monolayer thickness, it seems extremely unlikely that the insulating properties of metal/monolayer/metal sandwich structures are derived from the oxide layer(s) *alone* (Petty, 1990). One important experimental result that seems to support the view that tunnelling does occur in monomolecular LB films is the measurement of an *inelastic tunnelling spectrum* (Ginnai, 1982). Besides the normal elastic tunnelling of electrons (i.e., tunnelling through a potential barrier to a state at the same energy

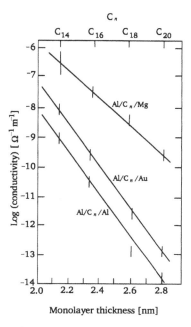

Figure 6.7 Logarithm of conductivity versus monolayer thickness for different metal/monolayer/metal structures. C_n is the number of carbon atoms in the monolayer material. (After Polymeropoulos, 1977. Reproduced with permission from the American Institute of Physics.)

level), it is possible for a small proportion, typically <1%, of the electrons to tunnel inelastically, losing energy to excite molecular vibrations in the barrier. The observation of this small number of relatively rare inelastic tunnelling events is convincing evidence for tunnelling through the LB layer.

6.4 Space-charge injection

If the electrical contacts to an insulating or semiconducting sample are Ohmic, these allow electron transfer between the electrodes and the sample and the resulting current is proportional to the applied voltage. Under certain conditions, however, the contacts can become super-Ohmic and the current is only limited by the *space-charge* between the electrodes. This conductivity regime is called *space-charge-limited* (Tredgold, 1966). In the simplest case, the current density J_{scl} varies with applied voltage V and sample thickness t as follows

$$J_{scl} \propto \frac{V^2}{t^3} \qquad (6.13)$$

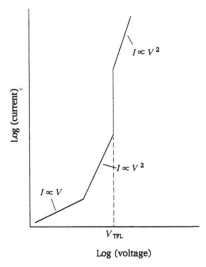

Figure 6.8 Space-charge-limited current versus voltage characteristics for an insulator containing traps. (V_{TFL} – trap-filled limit voltage.)

Figure 6.8 shows the expected current versus voltage behaviour under these conditions. The lowest voltage region of the curve corresponds to the situation in which the injection of excess carriers is negligible. At these voltages the volume conductivity dominates (i.e., Ohm's law). Only when the injected carrier density exceeds the volume generated carrier density will space-charge effects be observed and the quadratic current versus voltage dependence be observed.

If the insulator contains *traps* (localized centres which can capture electrons and/or holes), then the injected charge will fill these. When sufficient charge has been injected, the traps will become saturated. The voltage at which this occurs corresponds to the *trap-filled limit* and is shown as V_{TFL} in figure 6.8. Beyond V_{TFL} the insulator behaves as if it were trap-free and the quadratic current versus voltage relationship given in equation (6.13) is again observed.

Space-charge effects have been noted in several LB film systems. Electrical data for charge carrier flow normal to the plane of LB layers of a substituted anthracene are shown in figure 6.9 (Roberts et al., 1980a). A quadratic current versus voltage dependence is evident above 10 V. For some anthracene samples, a steeper dependence of current upon applied bias has been recorded; this was accompanied by blue light emission from the multilayer structure. The cause of this electroluminescence (section B.7) was assumed to be double carrier injection into the organic layer from the two electrodes (holes injected via an Au contact and electrons via an Al contact). Charge-injection

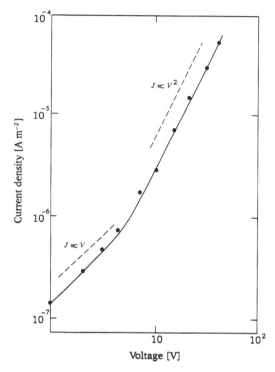

Figure 6.9 Current density normal to film plane versus voltage for a substituted anthracene LB film (51 layers) deposited onto an aluminium electrode. (After Roberts et al., 1980a. Reproduced with permission from Elsevier Science.)

effects have also been reported in multilayers of some charge-transfer complexes (Pearson et al., 1992).

6.5 Schottky and Poole–Frenkel effects

When an insulator (or semiconductor) is placed in contact with a metal, a re-distribution of charge will occur in the interface region (to minimize the free energy of the system). This will produce a distortion or bending of the energy bands. Abrupt changes in the potential energy do not occur however, as these would imply infinite electric fields. The potential step changes smoothly because of the *image force* effect. This arises because the metal's surface becomes positively charged by an escaping electron. The resulting potential barrier at the metal/insulator interface is called a *Schottky barrier*, figure 6.10(a). The electrons in the metal are filled to an energy called the *Fermi level* E_F and they must gain enough energy to overcome the barrier ϕ_B to get into the conduction band of the insulator. This process is called *Schottky*

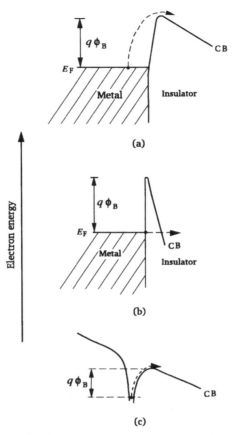

Figure 6.10 (a) Schottky emission of an electron from the Fermi level E_F in a metal into the conduction band (CB) of a semiconductor (or insulator); (b) Fowler–Nordheim tunnelling; (c) Poole–Frenkel effect.

emission and has a current density versus electric field dependence (Roberts et al., 1980b)

$$J \propto \exp\left(\frac{\beta E^{0.5}}{kT}\right) \tag{6.14}$$

where β is a constant and T is the temperature.

If the applied electric field is large enough, the barrier to electrons at the interface becomes very thin and electrons can tunnel directly from the metal to the conduction band of the insulator. This process is illustrated in figure 6.10(b) and is called *Fowler–Nordheim tunnelling*. The current density versus voltage behaviour is of the form (Simmons, 1971)

$$J \propto E^2 \exp\left(-\frac{\gamma}{E}\right) \tag{6.15}$$

where γ is another constant. This process can be thermally assisted, the electrons being excited to an energy state where they are able to tunnel through the barrier.

A similar process to Schottky emission can take place at impurity centres in the bulk of an insulator. This is the *Poole–Frenkel* effect and is illustrated in figure 6.10(c). The application of a high electric field will result in the lowering of the potential barrier around the impurity, allowing a carrier to escape into the conduction band of the insulator. The current versus voltage behaviour for Poole–Frenkel conduction has the same form as equation (6.14) for Schottky emission (although the β coefficient is different) and experimentally it can be difficult to distinguish between the two processes (Roberts et al., 1980b). However, Poole–Frenkel conduction is essentially a bulk effect and should show little dependence upon electrodes or upon the polarity of the field.

Both Poole–Frenkel and Schottky emission processes have been reported at high fields ($>10^7 \, \mathrm{V \, m^{-1}}$) in LB films of fatty acids and their salts. Other current versus voltage relationships have also been observed, e.g., $\log(\text{current}) \propto \text{voltage}^{0.25}$ (Petty, 1990). No clear picture has emerged from the many investigations that have taken place about which effect dominates under particular conditions. This suggests that the *high field d.c. electrical conductivity of these fatty acid LB film samples is predominantly dependent upon the preparation conditions.*

6.6 Dielectric processes

6.6.1 Capacitance

The real part of the relative permittivity of LB layers is usually found by measuring the capacitance of a metal/LB film/metal sandwich structure. The results of such an experiment have already been shown in chapter 3, figure 3.4. Problems of electrode and oxide resistance can complicate the interpretation of such data. If the electrode resistances can be neglected, the total measured capacitance C_M of the structure can be related to the permittivity of the LB film by

$$\frac{1}{C_M} = \frac{1}{A\epsilon_0}\left(\frac{Nt_{LB}}{\epsilon'_{rLB}} + \frac{t_{OX}}{\epsilon'_{rOX}}\right) \tag{6.16}$$

where the relative permittivity (real part) and thickness of the oxide layer are given by ϵ'_{rOX} and t_{OX}, respectively; ϵ'_{rLB} and t_{LB} represent the corresponding values for one LB monolayer; N is the total number of monolayers; and A is the electrode area. Equation (6.16) shows that the slope of a straight line plot of C_M^{-1} versus N gives the dielectric thickness (t_{LB}/ϵ'_{rLB}) for each monolayer while the intercept yields similar information about the oxide layer(s) on

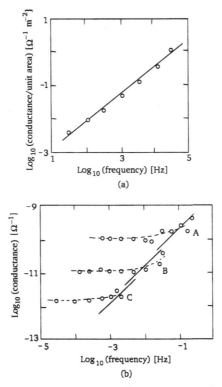

Figure 6.11 (a) Room temperature conductance versus frequency for 7-layer cadmium octadecanoate LB film. (b) Conductance at 77 K for 7-layer assemblies of cadmium hexadecanoate (A), cadmium octadecanoate (B) and cadmium eicosanoate (C). (After Sugi et al., 1979. Reproduced with permission from Gordon and Breach.)

the metallic electrode(s). This intercept will depend on the nature of the metal electrodes. The oxide associated with an aluminium electrode can be several nanometres in thickness, while a 'negligible' interfacial layer has been associated with magnesium electrodes (Geddes et al., 1989).

The real parts of the relative permittivity of LB films of fatty acids and fatty acid salts are usually in the range 2.0 to 3.0; e.g., $\epsilon_r' = 2.7$ for cadmium octadecanoate and $\epsilon_r' = 2.5$ for cadmium eicosanoate over a wide range of frequencies (Petty, 1990). The higher value in the former compound can be attributed to the stronger influence of the highly polarizable carboxyl groups in the shorter chain length material.

6.6.2 AC conductivity

Once electrode effects have been minimized (section 6.2.2), the power law relationship given by equation (6.7) between LB film a.c. conductivity and frequency is often observed (Petty, 1990). Figure 6.11(a) shows data taken for a 7-layer cadmium octadecanoate film (Sugi et al., 1979). The value of the exponent n is approximately unity here; however, values in the range 0.6–1.0 have been reported by various workers for fatty acid LB multilayers (Petty, 1990). At low frequencies, the *conductance* (the a.c. equivalent of resistance) of LB film assemblies becomes frequency independent, to agree with equation (6.5). Figure 6.11(b) shows this low frequency behaviour for different chain length fatty acid salt films (Sugi et al., 1979). The d.c. conductivity is seen to increase by approximately one order of magnitude as the length of the aliphatic tail is reduced by two carbon atoms.

6.6.3 LB film/electrolyte systems

Occasionally the electrical impedance of LB film samples is measured using a liquid electrolyte as the top electrode. This is particularly important for the investigation and development of sensor devices based on monolayer and multilayer films of biological molecules (Stelzle et al., 1993; Howarth and Petty, 1994).

The conductance and capacitance of a metal/LB film/electrolyte system are determined by the Gouy–Chapman layer (section 2.6.1) at the LB film/electrolyte interface, the LB film itself (including defects) and the electrolyte resistance. If the capacitance of the Gouy–Chapman layer is much larger than that of the LB film (as often happens), its contribution can be neglected. The presence of defects in the LB film will give rise to a parasitic defect capacitance, in parallel with the pure multilayer capacitance, and to a decrease in the film resistance.

Figure 6.12 shows a plot of the modulus of the impedance $|Z|$ as a function of frequency for a bilayer produced by vesicle (figure 2.6) fusion on a solid support (Stelzle et al., 1993); the points are experimental while the full curve is the best fit obtained by modelling the network with the resistor/capacitor combinations shown on the right of the figure. The electrical network of even this simple system is complicated, with contributions due to the bilayer capacitance and resistance (C_{LB} and R_{LB}), the electrolyte resistance (R_{El}) and defects in the film (C_D and R_D).

6.6.4 Permanent polarization

Polarization associated with monolayer and multilayer films can be measured directly with a probe technique (section 3.7.5). Surface potentials up to a few volts have been measured for various X-type, Y-type and alternate-layer fatty acid assemblies (Petty, 1990). These results could reflect the alignment of dipoles in the LB films; however, the potentials may also suggest the presence of bulk or surface charges. The polarization can also vary with temperature, giving rise to the phenomenon of *pyroelectricity*. If a pyroelectric sample is

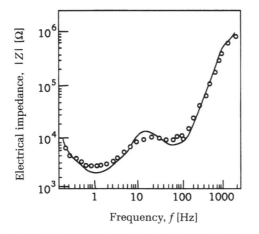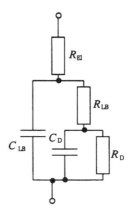

Figure 6.12 Modulus of electrical impedance |Z| versus frequency for a solid support/bilayer/electrolyte system produced by vesicle fusion. Points are experimental and the full curve corresponds to a best fit obtained with the equivalent circuit model shown on the right. R_{LB} and C_{LB} are due to the bilayer; R_{El} is associated with the electrolyte resistance; R_D and C_D result from defects in the organic layer. (Reprinted with permission from Stelzle, M., Weismüller, G. & Sackmann, E. (1993) J. Phys. Chem., 97, 2974–81. Copyright 1992 American Chemical Society.)

heated or cooled at a constant rate, a current I may be measured (Jones et al., 1988a), given by

$$I = pA \frac{dT}{dT} \tag{6.17}$$

where p is the pyroelectric coefficient and A is the sample area. Values of p for organic materials can be of the order of tens of $\mu C\,m^{-2}\,K^{-1}$ (Roberts, 1990). If a sample of area $10^{-4}\,m^2$ is heated at a rate of $1\,K\,s^{-1}$, the resulting current will be in the nanoampere range, easily measurable using modern electronic equipment.

Figure 6.13 shows the pyroelectric current obtained on heating and cooling an LB assembly consisting of a long-chain fatty acid alternated with a long-chain fatty amine (Jones et al., 1988a). As shown in chapter 4, figure 4.1, the deposition will result in a proton transfer from the acid head group to that of the amine, giving rise to an overall polarization component perpendicular to the multilayer plane. Many LB materials deposited as alternate-layer films or as X- or Z-type layers exhibit such behaviour (Roberts, 1990), including both preformed polymers and monolayers that are polymerized after LB deposition (Petty et al., 1992a & b; Tsibouklis et al., 1993). Some phospholipid alternate-layer films also show a pyroelectric response (Petty et al., 1992c). However, the largest pyroelectric coefficients are associated

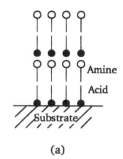

(a)

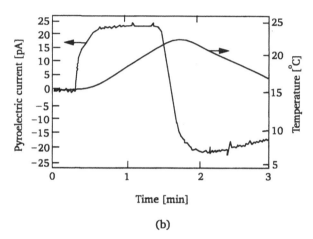

Time [min]

(b)

Figure 6.13 (a) Alternate-layer acid–amine LB assembly. (b) Time variation of temperature and pyroelectric current for a 99-layer acid–amine LB film. The arrows show which axis should be consulted for each curve. (After Jones et al., 1988a. © 1988 IEEE.)

with the simple (and perhaps more structurally ordered) acid/amine films. Noncentrosymmetric LB samples, such as those discussed above, are also expected to be *piezoelectric*, i.e., they will become polarized if they are subjected to a stress (Roberts, 1990). Such structures may exhibit nonlinear optical and electrooptical phenomena; these are discussed in chapter 7.

The polarization associated with a dielectric sample can be observed by heating the film under an applied electric field and observing the current. This experiment is called thermally stimulated conductivity (section 2.6) (Harrop, 1972; Blythe, 1979). Current peaks in the plot of current versus temperature will occur as the various types of dipoles revert to their random configuration. The experiment is relatively straightforward. However, detailed interpretation of the various current peaks in LB samples can be difficult (Jones et al., 1988b; Iwamoto and Sasaki, 1990). A variation on

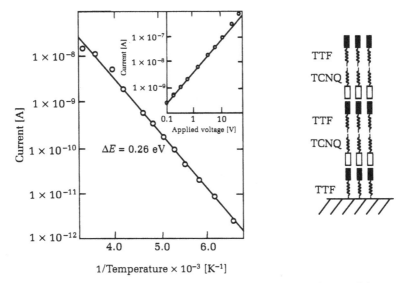

Figure 6.14 In-plane current versus reciprocal temperature for a 10-bilayer TTF/TCNQ alternate-layer structure deposited onto glass. The inset reveals the current versus voltage characteristics at room temperature. The multilayer structure is shown on the right. (After Pearson et al., 1989. Reproduced with permission from Elsevier Science.)

this method can be used to investigate the behaviour of a *floating* monolayer film during compression (Iwamoto et al., 1992). Here, the current that is measured results from the change in polarization at the monolayer/water interface as the molecules change their relative orientation.

6.7 Semiconducting multilayers

LB films of certain materials exhibit electrical in-plane conductivities that are many orders of magnitude greater than those observed for simple long-chain fatty acids. Examples are the porphyrins, phthalocyanines and charge-transfer complexes such as TCNQ and TTF (chapter 4) (Vandevyver, 1992). Sometimes, these 'active' materials are mixed with fatty acids as an aid to film deposition. Provided the proportion of the fatty acid is not too high (<30%), the in-plane conductivity of the mixed film will not be unduly affected. Figure 6.14 shows current versus voltage data (inset) and a plot of current (on a log scale) versus reciprocal temperature for an alternate-layer structure of amphiphilic TCNQ and TTF derivatives (Pearson et al., 1989). The linearity of the current versus voltage curve shows that the conductivity is Ohmic. Also, the conductivity value is independent of the electrode separation, suggesting that the effects of electrode resistance are negligible.

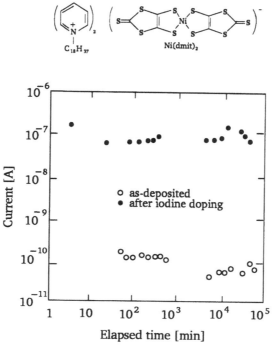

Figure 6.15 In-plane current (for an applied voltage of 10 V) versus time for a 19-layer film of a Ni(dmit)₂ complex (H₂dmit is the abbreviation for 4,5-dimercapto-1,3-dithiole-2-thione). The molecular formula is shown above. (After Pearson et al., 1992. Reproduced with permission from Elsevier Science.)

The room temperature conductivity is approximately $5 \times 10^{-1} \, \Omega^{-1} \, m^{-1}$ for the as-deposited film and shows no anisotropy in the film plane (i.e., the conductivity measured parallel to the dipping direction is the same as that perpendicular to the dipping direction). This measurement uses the total thickness of the LB film (i.e., the charge-transfer units plus the insulating aliphatic chains). Clearly, a higher conductivity value will be associated with the conducting planes. Over the temperature range studied, the LB film possesses a thermally activated conductivity of a form that can be obtained using equation (6.4), with $\Delta E = 0.26 \pm 0.01 \, eV$, similar to values observed in many inorganic semiconductors.

Relatively high in-plane conductivities of LB layers of amphiphilic charge-transfer complexes are sometimes achieved after doping. In this way, conductivities of 10^2–$10^3 \, \Omega^{-1} \, m^{-1}$ have been obtained. Figure 6.15 shows the effect of doping an LB film of an organometallic complex with iodine vapour (Pearson et al., 1992). The high conductivity of the LB film is retained for many weeks after the iodine treatment. Such stability is not always

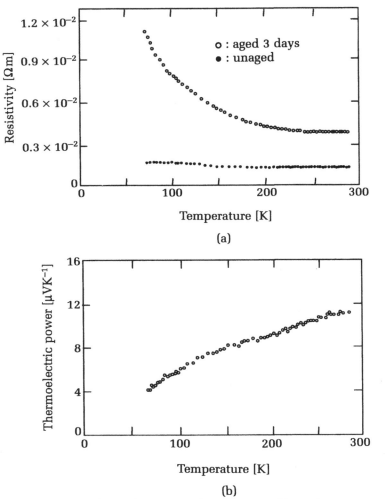

Figure 6.16 (a) Electrical resistivity of Au(dmit)$_2$ LB film in the temperature range 300–77 K. (b) Thermoelectric power of same film. (After
Miura et al., 1992. Reproduced with permission from Elsevier Science.)

observed in other materials and the conductivity decreases as the dopant
molecules diffuse out of the multilayer film.

An alternative method for doping films of charge-transfer complexes is to
use electrochemistry techniques (Miura et al., 1992; Goldenberg et al., 1994).
Although this approach seems satisfactory for a small number of LB layers,
the alkyl regions in thick films may hinder the diffusion of dopant ions.

LB film samples are not single crystals (chapter 5). The measured in-plane
conductivity values will therefore be average values that represent electrical conduction both within individual crystalline domains and between the domains.

Measurement of the *thermoelectric power* of a sample does not directly involve charge transport and may provide a better indication of conductivity within the individual domains. The method requires establishing a small temperature difference across the sample and monitoring the voltage generated (an effect familiar to anyone who has used a thermocouple). Figure 6.16 compares the electrical resistivity and thermoelectric power of an organometallic LB film, which has been rendered conducting by electrochemical doping (Miura et al., 1992). The thermoelectric power (figure 6.16(b)) is in the range 10–15 μV K^{-1} at room temperature and stable with time. It decreases approximately linearly with temperature down to 77 K. In contrast, the resistance changes with time (figure 6.16(a)). Immediately after doping, a slight metallic behaviour (resistance decreasing with temperature down to 200 K) is noted. After ageing, the film exhibits semiconducting characteristics. These results imply that there may be more than one process controlling the electrical behaviour of these LB layers: one *within* the domains and a different mechanism *between* them.

6.8 LB film devices

The semiconducting nature of some LB films suggests possible uses in electronic devices. However, it should be remembered that the conductivities are generally *anisotropic*. For example, the relatively high conductivities of the materials discussed above are *in-plane* values. Between the polar planes are layers of electrically insulating hydrocarbon chains. If a high conductivity *perpendicular* to the substrate plane is required, then the aliphatic chains must be either eliminated or 'short-circuited'. The former approach requires the deposition of molecules with very short or no aliphatic tails. Some success has been achieved using phthalocyanine or porphyrin compounds (section 4.4). Unfortunately, the resulting multilayer structures are generally more disordered than those built up from compounds substituted with long chains.

An alternative method is to develop molecular wires, based on unsaturated hydrocarbon chains (section 4.8.2) and to mix these with other monolayer-forming compounds. The relatively high conductivity along the chains of such molecules should enable electrical connection between the semiconducting layers in amphiphilic charge-transfer films to be established. The use of compounds based on β-carotene for this purpose has been attracting some interest. However, the orientation of the molecules in the multilayer assembly has not been firmly established. These materials are also rather unstable (Oshnishi et al., 1978; Wegmann et al., 1989; Williams et al., 1993).

6.8.1 Gas sensors

A much-studied electronic device using an LB layer is the gas sensor. This exploits the change in resistance of a semiconducting LB film on exposure

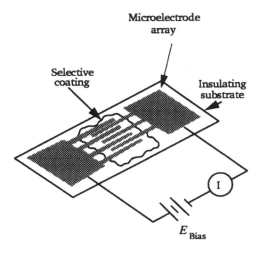

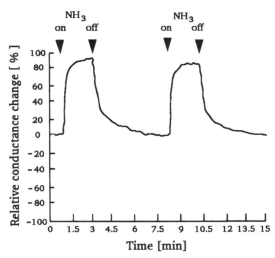

Figure 6.17 Response of a 45-layer copper phthalocyanine LB film to pulses of 2 p.p.m. ammonia. The LB film is deposited onto a planar interdigitated electrode array, as shown above. (After Wohltjen et al., 1985. © 1988 IEEE.)

to oxidizing or reducing gases (Roberts, 1990; Petty, 1992). Figure 6.17 shows an example of a copper phthalocyanine LB film, in an interdigitated electrode array, being used to sense small quantities of ammonia gas (Wohltjen et al., 1985). The sensitivity and selectivity of such sensors can be varied by changing the central metal atom and substitutions to the periphery of the phthalocyanine molecule. In this way sensitivities down to a few parts per million are readily achieved (Petty, 1992; Zhu et al., 1993; Cole et al., 1993).

The selectivity may be further enhanced by using an *array* of sensing elements, each with a slightly different response, and to exploit signal processing methods based on pattern recognition or artificial neural networks (Barker et al., 1993).

6.8.2 Diode structures

Semiconducting LB films have been used similarly to inorganic semi-conductors (e.g., Si, GaAs) in metal/semiconductor/metal structures. Perhaps the simplest example is that of a *diode*. Here, the LB film is sandwiched between metals of different *work functions* (the energy required to remove an electron from the Fermi level in the material to the vacuum level). In the ideal case, an *n-type* semiconductor (one in which the conductivity is dominated by electrons) should make an Ohmic contact to a low work function metal and a Schottky barrier (rectifying contact) to a high work function metal (Rhoderick, 1978). The reverse will be true for a *p-type* semiconductor (conductivity dominated by holes).

Figure 6.18 shows an example of a Schottky barrier formed by sandwiching a phthalocyanine LB film between aluminium and indium–tin-oxide electrodes (Hua et al., 1987a). The LB material is a *p*-type semiconductor and the aluminium/LB film interface provides the rectifying junction. The current versus voltage characteristics of this device, figure 6.19, are asymmetrical, exhibiting significant rectification. Under illumination the structure acts as a photovoltaic cell. Electrons and holes generated in, and close to, the barrier region will be moved in opposite directions by the high electric field associated with the Schottky contact. This results in the dissipation of power in an external circuit. Although the voltages produced by such devices are of the same order as those from silicon photocells, the currents are very many orders of magnitude less.

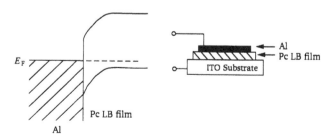

Figure 6.18 Energy band structure of a rectifying contact between an unsubstituted phthalocyanine (Pc) LB film and an aluminium contact. The electrical contact between the indium-tin-oxide (ITO) substrate and the LB film is Ohmic.

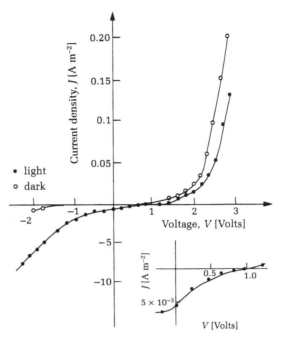

Figure 6.19 Current density versus voltage characteristics in the dark (○) and light (●) for an indium-tin-oxide/substituted phthalocyanine LB film/ aluminium structure. Inset: the light characteristics in an expanded form. (After Hua et al., 1987a. Reproduced with permission from Elsevier Science.)

Figure 6.20 Donor/tunnel bridge/acceptor rectifier molecule. (After Aviram and Ratner, 1974.)

Of particular interest is the possibility of observing *molecular rectification* using monolayer or multilayer films. This follows the prediction of Aviram and Ratner (1974) that an asymmetric organic molecule containing a donor and an acceptor group separated by a short σ-bonded bridge (allowing tunnel- ling), as shown in figure 6.20, should exhibit diode characteristics. There have been many recent attempts to obtain this effect in LB systems (Metzger and Panetta, 1991; Martin et al., 1993). Asymmetric current versus voltage behav- iour has certainly been recorded for many LB film metal/insulator/metal (MIM) structures (e.g., figure 6.19). However, the results are often open to several interpretations. For example, the current versus voltage characteristics for tunnelling between different metals should exhibit some asymmetry

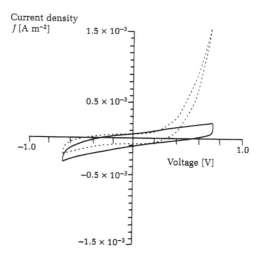

Figure 6.21 Current density versus voltage curve from a 7-layer LB film of a zwitterionic compound sandwiched between Ag and Mg electrodes. Full line: bleached material; dashed line: unbleached material. Reprinted (abstracted) with permission from Martin, A. S., Sambles, J. R. & Ashwell, G. J. (1993) Physical Review Letters, **70**, *218–21. Copyright 1993 The American Physical Society.*

(Simmons, 1971). Furthermore, Schottky barriers can exist at metal/semiconductor and metal/insulator/semiconductor interfaces. Figure 6.21 shows an interesting result of this type (Martin et al., 1993). The current versus voltage behaviour was obtained for an Ag/seven Z-type monolayers of a zwitterionic compound/Mg junction. The rectifying characteristics (dashed line) disappear (full line) for a bleached LB film (formed by deposition from a subphase containing low levels of metal ions). This suggests that the rectification is directly associated with the organic thin film, rather than, say, a result of the difference in the work functions of the two metal electrodes. In other work on diode structures, single electron tunnelling events were observed which might be related to the charging of single molecules or molecular stacks (Fischer et al., 1994).

The ultimate test for a molecular rectifier of the type suggested by Aviram and Ratner would be to produce complementary structures, in which the LB molecules have been deposited in different orientations (i.e., tail-down and head-down); the reversal of the diode action would then show that the electrical behaviour originated from a molecular process, rather than from a Schottky barrier effect.

6.8.3 LB layers on inorganic semiconductors

There have been many studies of metal/LB film insulator/semiconductor (MIS) structures (Roberts, 1990). Mostly, the LB film has been a fatty acid

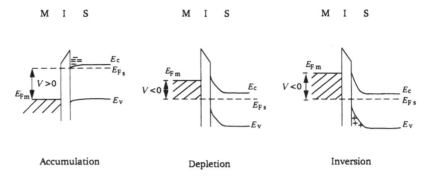

Figure 6.22 Energy band structure for a metal/insulator/semiconductor (MIS) device for different applied biases, V. (E$_c$ – conduction band edge; E$_v$ – valence band edge; E$_{Fm}$ – metal Fermi level; E$_{Fs}$ – semiconductor Fermi level.)

or a polymer. The role of the LB film is essentially a *passive* one, providing an insulating layer across which a voltage can be sustained, 'decoupling' the Fermi level of the semiconductor from that of the metal top electrode.

Figure 6.22 shows schematic diagrams of the energy band structures of simple MIS devices with various bias voltages applied to the top metal contact. In this example the semiconductor is *n*-type and the Fermi level is closer to the edge of the conduction band than to the valence band. There are essentially three cases that can exist at the semiconductor surface (Sze, 1969). When a positive voltage ($V > 0$) is applied to the metal plate, the top of the conduction band in the semiconductor bends downwards and closer to the Fermi level E_{Fs}. The valence band is bent down in a similar way. For an ideal MIS device, there is no current flow in the structure and the Fermi level remains constant in the semiconductor. The carrier density in the conduction band depends exponentially on the energy difference between the conduction band edge E_c and E_{Fs} (Sze, 1969). Therefore, the band bending causes an accumulation of majority carriers (electrons) near the semiconductor surface. This regime is called *accumulation*. When a small negative voltage ($V < 0$) is applied, the bands bend upwards, and the majority carriers are depleted near the semiconductor surface; this is *depletion*. Finally, if a larger negative voltage is applied, E_{Fs} becomes closer to the valence band at the semiconductor surface and minority carriers (holes) are generated; this is called *inversion*. Similar effects occur with a *p*-type semiconductor, but the polarity of the applied voltage is reversed.

The different regimes in an MIS structure may be conveniently monitored by measuring the *differential capacitance* (i.e., $\partial C/\partial V$) of the device. Figure 6.23 shows the capacitance and conductance at 100 kHz as a function of bias for a metal/cadmium octadecanoate/*p*-type CdTe MIS structure (Petty and

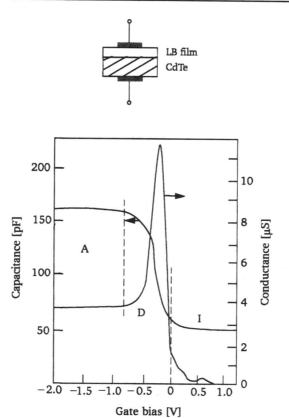

Figure 6.23 Capacitance and conductance versus voltage characteristics at 100 kHz for a Au/cadmium octadecanoate LB film/p-type CdTe MIS structure. The arrows show which axis should be consulted for each curve. (A – accumulation; D – depletion; I – inversion.) (After Petty and Roberts, 1979. Reproduced with permission from the IEE.)

Roberts, 1979; Roberts, 1990). The accumulation (A), depletion (D) and inversion (I) regimes are marked on the figure. In accumulation (negative bias for the *p*-type CdTe), the constant capacitance is simply due to capacitance of the LB film (the device can be thought of as a parallel plate capacitor with the LB film as the dielectric). The decrease in the capacitance in the depletion region is due to the capacitance of the depletion layer (which decreases with increasing negative bias) appearing in series with the LB film capacitance. When minority carriers (electrons here) are generated in the inversion regime, the applied electric field cannot further penetrate the depletion region, which remains a constant thickness. The capacitance therefore remains constant, as shown in the figure. This is the case at high frequencies, at very low frequencies, the capacitance in inversion would be expected to rise to the

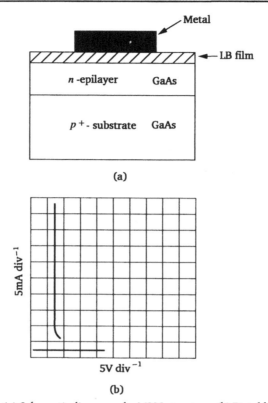

Figure 6.24 (a) Schematic diagram of a MISS structure. (b) Bistable switching for an MISS device based on GaAs and a 22-tricosenoic acid LB film. (After Thomas et al., 1984. Reproduced with permission from the IEE.)

value in accumulation, as the minority carriers respond to the a.c. measuring voltage (Sze, 1969). The large conductance peak in the depletion region in figure 6.23 is due to losses occurring as interface traps fill and empty. Similar capacitance and conductance characteristics have been noted for a wide range of LB film/inorganic semiconductor combinations (Roberts, 1990).

If the insulating layer in an MIS structure is made thin enough, a d.c. current can pass between the top metal layer and the semiconductor (by quantum mechanical tunnelling or via defects in the film). The insulating layer in such a device can be used to control the carrier transport and to influence the degree of band bending (Schottky barrier height) in the semiconductor. In this way MIS structures with enhanced photovoltaic and/or electroluminescent properties can be formed (Pulfrey, 1978; Card and Rhoderick, 1973). The obvious candidates as the insulators for most semiconductors (e.g., silicon) are their natural oxide layers, which can be grown to a specified thickness. Unfortunately, not all semiconductors possess a high quality native oxide. In these cases LB films have been used to improve the efficiencies of both electroluminescent and photovoltaic devices (Petty et al., 1985; Roberts, 1990). The insertion of a phthalocyanine LB film between an Au top electrode and a ZnSe substrate has resulted in the observation of blue electroluminescence (Hua et al., 1987b).

A more complex MIS device incorporating an LB layer is shown in figure 6.24; this has two layers of semiconductor, each with a different doping level. The resulting MISS structure exhibits bistable switching characteristics (Thomas et al., 1984; Roberts, 1990). Potential applications include memories, shift registers and light-sensitive or gas-sensitive switches.

6.8.4   LB film FET

The ability to control the surface charge on a semiconductor using an MIS structure incorporating an *insulating LB film* allows a *field effect transistor* (FET) to be fabricated. Transistor action has been demonstrated in a few fatty acid LB film structures (Petty, 1992). However, a problem with all work of this type has been the relatively low melting point of the organic film. An alternative is to exploit a polymeric material. Larkins et al. (1989) have described the fabrication of silicon FETs using an amphiphilic diacetylene (section 4.7.1). Following polymerization of the LB film, the electrical characteristics were found to be highly stable.

An alternative approach is to use the LB film as the *semiconducting layer* in an FET device. Here, the LB film would need to possess a reasonable in-plane conductivity: suitable candidates are amphiphilic charge-transfer compounds and conductive polymers (sections 4.6 and 4.7). Figure 6.25 shows two alternative structures for this thin film transistor (Sze, 1969). The FET requires three electrical connections: *source, drain* and *gate*. The source and drain connections are made to the semiconducting layer and the gate connection is to the insulator. Figure 6.25(a) shows a conventional approach. The metal source and drain connections are first deposited onto an insulating substrate. These are overlaid by the semiconducting film and the insulator is deposited on top; finally, the gate contact is made. This arrangement is generally not favoured by LB workers as the LB film must withstand post deposition chemical and thermal processing. In the arrangement shown in figure 6.25(b) the semiconducting film is deposited in the final stage of the FET fabrication. LB film FET devices using conductive polymers and charge-transfer complexes have both been described (Paloheimo et al., 1992; Pearson et al., 1994). Figure 6.26 shows a set of current versus voltage characteristics for a structure incorporating an amphiphilic organometallic complex, similar to that discussed in section 6.7 (Pearson et al., 1994). For high voltages applied between the source and drain V_{DS} and for large gate biases V_G, the drain current I_D shows saturation. This agrees with standard

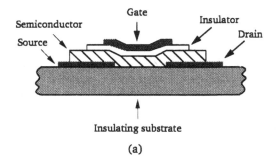

(a)

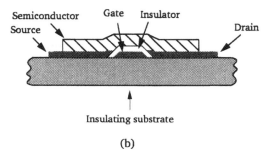

(b)

Figure 6.25 Alternative structures for a thin film field effect transistor.

thin film transistor theory, which predicts the saturation value I_{Dsat} as (Sze, 1969)

$$I_{\text{Dsat}} = \left(\frac{w}{2l}\right) \frac{C_{\text{G}}}{A} \mu (V_{\text{G}} - V_0)^2 \tag{6.18}$$

where w and l are the channel width and length, respectively, C_{G}/A is the gate capacitance per unit area, μ the carrier mobility and V_0 is a constant (called the *pinch-off voltage*). As expected, the carrier mobilities for both the organo-metallic and polymer LB film FETs are quite small ($\approx 10^{-9}\,\text{m}^2\,\text{V}^{-1}\,\text{s}^{-1}$ and 10^{-9}–$10^{-12}\,\text{m}^2\,\text{V}^{-1}\,\text{s}^{-1}$, respectively, cf. values of 10^{-2}–$10^{-1}\,\text{m}^2\,\text{V}^{-1}\,\text{s}^{-1}$ for FETs based on single crystal silicon). However, the mobility for the organometallic device can be increased by several orders of magnitude (to around $10^{-5}\,\text{m}^2\,\text{V}^{-1}\,\text{s}^{-1}$) by doping the LB layer.

6.8.5 Future developments

Once reliable and reproducible electrical data can be obtained for LB film MIM or MIS systems, more complex applications have been suggested. As an example, Burrows et al. (1989) have described the principle of a three-dimensional molecular memory based on LB films. The device requires a

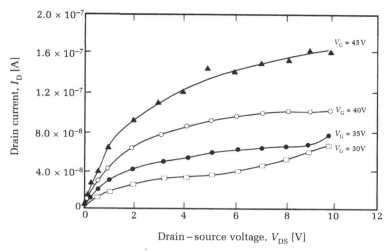

Figure 6.26 Drain current I_D versus drain–source voltage V_{DS} at various gate voltages V_G for a thin film FET incorporating an LB film of an amphiphilic organometallic charge-transfer compound. (After Pearson et al., 1994. Reproduced with permission from Elsevier Science.)

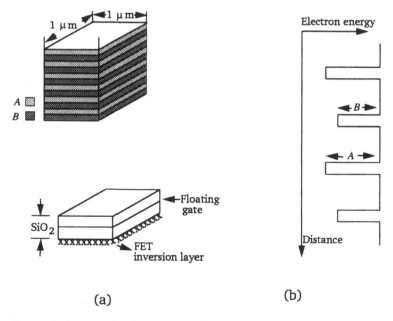

(a)   (b)

Figure 6.27 (a) Molecular memory holding N bits (top) compared to a conventional silicon memory holding 1 bit (bottom). (b) Electron energy versus perpendicular distance for molecular memory with no applied electric field. (After Burrows et al., 1989.)

molecule with a central conjugated region of high *electron affinity* (for an *n*-type material, the electron affinity is the energy difference between the bottom of the conduction band and the vacuum level) surrounded by aliphatic substituents of low electron affinity. A multilayer structure, as shown in figure 6.27, could be used to store one *N*-bit word, the presence or absence of charge on the *n*th layer representing a 0 or 1 of the *n*th bit. The LB film could be assembled on the gate of an FET and, on application of an electric field, transport of bits across the layers may be detected as induced charge on the gate. A modification to the basic structure circumvents problems due to the dispersion of the charge packet as it crosses the layers (Burrows, 1989). Alternative memory devices utilize the tip of a scanning tunnelling microscope (STM, see section 5.6) to address the various memory elements (Sakai et al., 1988).]

References

Aviram, A. and Ratner, M. A. (1974) Molecular rectifiers, *Chem. Phys. Lett.*, **29**, 277–83

Barker, P., Chen, J. R., Agbor, N. E., Monkman, A. P., Mars, P. and Petty, M. C. (1993) Vapour recognition using organic films and artificial neural networks, *Sensors and Actuators B*, **17**, 143–7

Blythe, A. R. (1979) *Electrical Properties of Polymers*, Cambridge University Press, Cambridge

Burrows, P. E. (1989) Electron transport in Langmuir–Blodgett films, PhD thesis, University of London, UK

Burrows, P. E., Donovan, K. J. and Wilson, E. G. (1989) Electron motion perpendicular to Langmuir–Blodgett multilayers of conjugated macrocyclic compounds: the organic quantum well, *Thin Solid Films*, **179**, 129–36

Card, H. C. and Rhoderick, E. H. (1973) The effect of an interfacial layer on minority carrier injection in forward-biased silicon Schottky diodes, *Solid State Electron.*, **16**, 365–74

Cole, A., McIlroy, R. J., Thorpe, S. C., Cook, M. J., McMurdo, J. and Ray, A. K. (1993) Substituted phthalocyanine gas sensors, *Sensors and Actuators B*, **13/14**, 416–19

Ferraro, J. R. and Williams, J. M. (1987) *Introduction to Synthetic Electrical Conductors*, Academic Press, Orlando

Fischer, C. M., Burghard, M., Roth, S. and v. Klitzing, K. (1994) Organic quantum wells: molecular rectification and single electron tunnelling, *Europhysics Letts.*, **28**, 129–34

Geddes, N. J., Parker, W. G., Sambles, J. R., Jarvis, D. J. and Couch, N. R. (1989) A metal (organic bilayer) metal capacitor, *Thin Solid Films*, **168**, 151–6

Ginnai, T. M. (1982) An investigation of electron tunnelling and conduction in Langmuir films, PhD thesis, Leicester Polytechnic, UK

Goldenberg, L. M., Cooke, G., Pearson, C., Monkman, A. P., Bryce, M. R. and Petty, M. C. (1994) Electrochemical properties of hexadecanoyltetrathiafulvalene Langmuir–Blodgett films, *Thin Solid Films*, **238**, 280–4

Harrop, P. J. (1972) *Dielectrics*, Butterworths, London

Honig, E. P. and de Koning, B. R. (1978) Transient and alternating electric currents in thin organic films, *J. Phys. C.*, **11**, 3259–71

Howarth, V. A. and Petty, M. C. (1994) Impedance spectroscopy of biomembrane Langmuir–Blodgett films, *Thin Solid Films*, **244**, 951–4

Hua, Y. L., Petty, M. C., Roberts, G. G., Ahmad, M. M., Hanack, M. and Rein, M. (1987a) Photoelectric properties of substituted silicon phthalocyanine Langmuir–Blodgett film Schottky barrier and metal/insulator/semiconductor devices, *Thin Solid Films*, **149**, 163–70

Hua, Y. L., Petty, M. C., Roberts, G. G., Ahmad, M. M., Yates, N. M., Maung, N. J. and Williams, J. O. (1987b) Blue electroluminescence from ZnSe/Langmuir–Blodgett film MIS diodes, *Electron. Lett.*, **23**, 231–2

Iwamoto, M. and Sasaki, T. (1990) Thermally stimulated discharge of Au/LB/air-gap/Au structures incorporating cadmium arachidate Langmuir–Blodgett films, *Jap. J. Appl. Phys.*, **29**, 536–9

Iwamoto, M., Majima, Y. and Watanabe, A. (1992) Detection of the reorganization of monolayers at the air–water interface by displacement current measurement, *Thin Solid Films*, **210/211**, 101–4

Jones, C. A., Petty, M. C. and Roberts, G. G. (1988a) Langmuir–Blodgett films: a new class of pyro-electric materials, *IEEE Trans. on Ultrasonics, Ferroelectrics and Frequency Control*, **35**, 736–40

Jones, C. A., Petty, M. C., Davies, G. and Yarwood, J. (1988b) Thermally stimulated discharge of alternate-layer Langmuir–Blodgett film structures, *J. Phys. D: Appl. Phys.*, **21**, 95–100

Jonscher, A. K. (1983) *Dielectric Relaxation in Solids*, Chelsea Dielectrics Press, London

Larkins, G. L., Jr., Fung, C. D. and Rickert, S. E. (1989) Integrated metal–Langmuir–semiconductor field effect transistors, *Thin Solid Films*, **180**, 217–25

Martin, A. S., Sambles, J. R. and Ashwell, G. J. (1993) Molecular rectifier, *Phys. Rev. Lett.*, **70**, 218–21

Metzger, R. M. and Panetta, C. A. (1991) The quest for unimolecular devices, *New J. Chem.*, **15**, 209–21

Miura, Y. F., Takenaga, M., Kasai, A., Nakamura, T., Nishio, Y., Matsumoto, M. and Kawabata, Y. (1992) Electronic transport in Langmuir–Blodgett films of metal-(dmit)$_2$, *Thin Solid Films*, **210/211**, 306–8

Okamoto, Y. and Brenner, W. (1964) *Organic Semiconductors*, Rheinhold, New York

Ohnishi, T., Hatakeyama, M., Yamamoto, N. and Tsubomura, H. (1978) Electrical and spectroscopic investigations of molecular layers of fatty acids including carotene, *Bull. Chem. Soc. Japan*, **51**, 1714–16

Paloheimo, J., Stubb, H., Yli-Lahti, P., Dyreklev, P. and Inganäs, O. (1992) Electronic and optical studies with Langmuir–Blodgett transistors, *Thin Solid Films*, **210/211**, 283–6

Pearson, C., Dhindsa, A. S., Bryce, M. R. and Petty, M. C. (1989) Alternate-layer Langmuir–Blodgett films of long-chain TCNQ and TTF derivatives, *Synthetic Metals*, **31**, 275–9

Pearson, C., Dhindsa, A. S., Petty, M. C. and Bryce, M. R. (1992) Electrical properties of Langmuir–Blodgett films of a Ni(dmit)$_2$ charge-transfer complex, *Thin Solid Films*, **210/211**, 257–60

Pearson, C., Moore, A. J., Petty, M. C. and Bryce, M. R. (1994) A field effect transistor based on Langmuir–Blodgett films of an Ni(dmit)$_2$ charge transfer complex, *Thin Solid Films*, **244**, 932–5

Petty, M. C. (1990) Characterization and properties, in *Langmuir–Blodgett Films*, ed. G. G. Roberts, pp. 133–221, Plenum Press, New York

Petty, M. C. (1992) Possible applications for Langmuir–Blodgett films, *Thin Solid Films*, **210/211**, 417–26

Petty, M. C. and Roberts, G. G. (1979) CdTe/Langmuir-film M.I.S. structures, *Electron. Lett.*, **15**, 335–6

Petty, M. C., Batey, J. and Roberts, G. G. (1985) A comparison of the photovoltaic and electrolumi-nescent effects in GaP/Langmuir–Blodgett film diodes, *IEE Proc.*, **132**, Pt. I, 133–9

Petty, M., Tsibouklis, J., Song, Y. P., Yarwood, J., Petty, M. C. and Feast, W. J. (1992a) 22-tricosenoic acid/1-docosylamine alternate-layer Langmuir–Blodgett films: polymerization, pyroelectric prop-erties and infrared spectroscopic studies, *J. Mater. Chem.*, **2**, 87–91

Petty, M., Tsibouklis, J., Davies, F., Hodge, P., Petty, M. C. and Feast, W. J. (1992b) Pyroelectric Langmuir–Blodgett films prepared using preformed polymers, *J. Phys. D: Appl. Phys.*, **25**, **1032–5**

Petty, M., Tsibouklis, J., Petty, M. C. and Feast, W. J. (1992c) Pyroelectric behaviour of synthetic biomembrane structures, *Thin Solid Films*, **210/211**, 320–3

Polymeropoulos, E. E. (1977) Electron tunnelling through fatty-acid monolayers, *J. Appl. Phys.*, **48**, 2404–7

Pulfrey, D. L. (1978) MIS solar cells: a review, *IEEE Trans.*, **ED-25**, 1308–17

Rhoderick, E. H. (1978) *Metal–Semiconductor Contacts*, Clarendon Press, Oxford

Roberts, G. G. (1990) Potential application of Langmuir–Blodgett films, in *Langmuir–Blodgett Films*, ed. G. G. Roberts, Plenum Press, New York

Roberts, G. G., McGinnity, T. M., Barlow, W. A. and Vincett, P. S. (1980a) A.C. and D.C. conduction in lightly substituted anthracene Langmuir films, *Thin Solid Films*, **68**, 223–32

Roberts, G. G., Apsley, N. and Munn, R. W. (1980b) Temperature dependent electronic conduction in semiconductors, *Phys. Reports*, **60**, 59–150

Sakai, K., Matsuda, H., Kawada, H., Eguchi, K. and Nakagiri, T. (1988) Switching and memory phenomena in Langmuir–Blodgett films, *Appl. Phys. Lett.*, **53**, 1274–6

Simmons, J. G. (1971) *DC Conduction in Thin Films*, Mills and Boon, London

Stelzle, M., Weissmüller, G. and Sackmann, E. (1993) On the application of supported bilayers as receptive layers for biosensors with electrical detection, *J. Phys. Chem.*, **97**, 2974–81

Sugi, M., Fukui, T. and Iizima, S. (1979) Structure-dependent feature of electron transport in Langmuir multilayer assemblies, *Mol. Cryst. Liq. Cryst.*, **50**, 183–200

Sze, S. M. (1969) *Physics of Semiconductor Devices*, Wiley, New York

Thomas, N. J., Petty, M. C., Roberts, G. G. and Hall, H. Y. (1984) GaAs/LB film MISS switching device, *Electron. Lett.*, **20**, 838–9

Tredgold, R. H. (1966) *Space Charge Conduction in Solids*, Elsevier, Amsterdam

Tsibouklis, J., Pearson, C., Song, Y. P., Warren, J., Petty, M., Yarwood, J., Petty, M. C. and Feast, W. J. (1993) Pentacosa-10,12-diynoic acid/heneicosa-2,4-diyneamine alternate-layer Langmuir–Blodgett films: synthesis, polymerization and electrical properties, *J. Mater. Chem.*, **3**, 97–104

Vandevyver, M. (1992) New trends and prospects in conducting Langmuir–Blodgett films, *Thin Solid Films*, **210/211**, 240–5

Wegmann, A., Tieke, B., Pfeiffer, J. and Hilti, B. (1989), Electrical conductivity of novel Langmuir–Blodgett films containing ethyl-β-apo-8'-carotenoate, an amphiphilic carotene, *J. Chem. Soc., Chem. Commun.*, 586–8

Williams, G., Bryce, M. R. and Petty, M. C. (1993) Electrical properties of multilayer films containing a carotene derivative, *Mol. Cryst. Liq. Cryst.*, **229**, 83–90

Wohltjen, H., Barger, W. R., Snow, A. W. and Jarvis, N. L. (1985) A vapour-sensitive chemiresistor fabricated with microelectrodes and a Langmuir–Blodgett organic semiconductor film, *IEEE Trans. Elec. Dev.*, **ED-32**, 1170–4

Zhu, D. G., Cui, D. F., Petty, M. C. and Harris, M. (1993) Gas sensing using Langmuir–Blodgett films of a ruthenium porphyrin, *Sensors and Actuators B*, **12**, 111–4

7

Optical properties

7.1 Refractive index

The refractive index of materials is determined by the interaction of electro-magnetic (EM) radiation with the molecules which they comprise (appendix B). This depends not only on the orientation of the electric field vector of the incident EM wave, but also on that of the electric dipoles produced in neighbouring molecules. Figure 7.1 shows an ideal arrangement of molecules in an LB monolayer. The sample coordinate system is (x, y, z) while that of the principal axes of the molecules is (x', y', z'). Careful measurements on fatty acid LB layers show that the films possess a biaxial symmetry with three independent permittivity values (Barnes and Sambles, 1987). However, two of the indices are very close in value and LB films are often approximated as uniaxial.

Several approaches can be used to measure the refractive indices of thin organic films (Petty, 1990). In some techniques, the film thickness is also obtained (section 5.8). The more popular methods, based on *ellipsometry*, *surface plasmon resonance* and *waveguiding* are discussed below. Figure 7.2 summarizes the results of such experiments by various workers using cadmium eicosanoate LB films (Swalen et al., 1978). It is evident that the refractive index for the extraordinary ray n_e (p-polarization) is greater than that for the ordinary ray n_o (s-polarization) by 0.04. Furthermore, both refractive indices increase as the wavelength is decreased. The refractive index of fatty acid salt films can be modified somewhat by controlling the salt:acid ratio in the film. This is accomplished either by varying the pH of the subphase during film deposition (section 2.5.1) (Pitt and Walpita, 1980) or by the process of skeletonization (Tomar, 1974). The latter involves soaking the deposited film in a suitable solvent to remove the free acid (section 3.9.1).

Larger refractive index values are usually noted for LB films of dye compounds. For example, the oligomers shown in Chapter 4, figure 4.15,

167

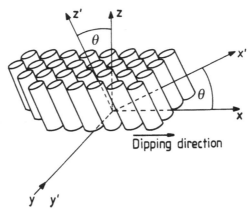

Figure 7.1 Ideal arrangement of rod-shaped molecules in a monolayer film. (After Barnes and Sambles, 1987.)

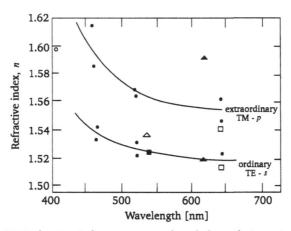

Figure 7.2 Refractive index versus wavelength for cadmium eicosanoate. ▲, Fleck, 1966; □, den Engelsen, 1971; ■, Drexhage, 1966; △, Blodgett, 1939; ○, Forster, 1966; ●, Swalen et al., 1978. The solid curves represent a smooth average. TM–transverse-magnetic (p-polarized); TE–transverse-electric (s-polarized) (see appendix B.1). (After Swalen et al., 1978. Reproduced with permission from Elsevier Science.)

possess values of $n_e \approx 1.8$ and $n_o \approx 1.6$ (Allen et al., 1993). The birefringence $(n_e - n_o)$ associated with these films is therefore significantly greater than that of fatty acid and fatty acid salt LB layers.

7.2 Ellipsometry

The reflectivity from a surface differs in both amplitude and phase for s- and p-polarized incident radiation (appendix B). *Ellipsometry* is a comparison of

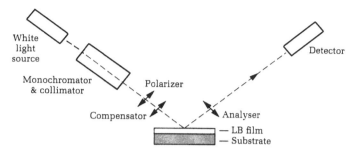

Figure 7.3 Optical system of an ellipsometer.

these reflectivities (Swalen, 1986). Figure 7.3 shows the optical arrangement of a typical measuring instrument. The presence of a surface film alters the ratio of the electric vectors vibrating in the plane of incidence and perpendicular to it, also their difference in phase Δ. In the former case, an angle ψ is defined as $\tan^{-1}$ (reflectivity amplitude ratio). The theory of ellipsometry correlates the parameters ψ and Δ with the optical thickness of the layer and the optical constants of the surface.

If there is no mixing between the polarizations the reflection behaviour can be fully described by two Fresnel coefficients, the amplitude reflectance for p-polarized input to p-polarized output $r_{\parallel}$ and the equivalent for s-polarization $r_{\perp}$ (appendix B).

$$\frac{r_{\parallel}}{r_{\perp}} = \tan \psi \, e^{j\Delta} \qquad (7.1)$$

It is usual to measure indirectly the ratio of these quantities by finding ψ and Δ. Using the Fresnel equations it is relatively straightforward to *calculate* ψ and Δ. However, to evaluate the refractive index and thickness of a thin film from measured values of ψ and Δ, the situation is more complicated. Computer programs do this in commercial ellipsometers.

A major problem with ellipsometric measurements of the optical constants of LB films concerns the nature of the thin film system under investigation. The LB layers are invariably deposited onto solid surfaces: these may be metals (or metallized glass microscope slides) or semiconductors. For such substrates, there is almost certainly a surface ('oxide') layer between the bulk substrate and the LB film. This may result from exposing an evaporated metal film or a freshly etched semiconductor to the atmosphere. It may also be augmented during the LB deposition process as the substrate is lowered into (or left for some time in) the aqueous subphase. Such an interfacial film may be several nanometres in thickness and can introduce considerable errors if its presence is ignored. The usual solution is to modify the optical constants of the underlying substrate by taking measurements on an uncoated portion of the substrate (Petty, 1990).

Table 7.1 *Ellipsometric angles and lossless isotropic fit for LB layers of 22-tricosenoic acid.*
(After Cresswell, 1992.)

Number of layers	Δ [degrees]	ψ	Refractive index	Average monolayer thickness [nm]
10	112.94	16.83	1.483	2.84
20	85.84	26.47	1.514	2.87
30	74.44	37.31	1.527	2.84
40	72.93	58.06	1.518	2.81
50	287.17	73.46	1.505	2.83
60	281.55	42.32	1.481	2.91

In most ellipsometric work on LB films it is assumed that the organic layers are nonabsorbing and isotropic. This is clearly not so in practice and the theory of ellipsometry has been extended to anisotropic films (Petty, 1990). Although the theoretical reflectances and transmittances of LB films depend on whether they are assumed to behave in an isotropic or in an anisotropic manner, the allowance for anisotropy does not appear to affect appreciably the ellipsometric thickness values for simple fatty acid layers. For example, using an anisotropic model, den Engelsen (1971) found that the thickness of a cadmium eicosanoate monolayer was 2.68 ± 0.02 nm, compared to a value of 2.70 ± 0.03 nm, obtained if the films were assumed to be optically isotropic. Table 7.1 shows the results of fitting data for 22-tricosenoic acid monolayers to a lossless isotropic model (Cresswell, 1992). The six sets of data were also fitted, independently, by a uniaxial model. This gave an ordinary index of 1.511 ± 0.001, an extra-ordinary index of 1.552 ± 0.001 and a layer thickness of 2.89 ± 0.01 nm. Finally, the data for all six regions were used together to give the best isotropic fit: this produced a refractive index of 1.5125 and a layer thickness of 2.89 nm. The form of these results is very similar to that reported by den Engelsen. Whichever model is used, the layer thickness varies only very slightly and the isotropic index is found to be close to the ordinary index of the uniaxial model.

7.3 The evanescent field

For light incident from a dense to a less dense medium at an angle of incidence greater than the critical angle θ_C, there will be total internal reflection (appendix B). This situation is sketched in figure 7.4. If the reflected power R is recorded as a function of the angle of incidence, R becomes unity at the critical angle. At angles greater than θ_C, the reflected beam has the same amplitude as the incident beam, but a phase difference between 0 and π. The effect on the transmitted beam is rather strange. This takes on the form of a wave that

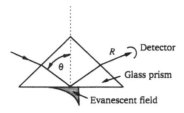

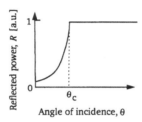

Figure 7.4 Total internal reflection of light at the base of a glass prism.
θ_C = critical angle.

travels along the boundary with an amplitude that decays perpendicular to the
boundary. The wave carries no energy away from the boundary and is called
an *evanescent wave*. Figure 7.5 shows how both the amplitude and intensity
(square of the amplitude) of a typical evanescent wave vary with distance from
a glass/air interface (Swalen, 1986). It is evident that the penetration depth
(into the air medium in this example) is similar to the wavelength of light.
The electric field associated with evanescent waves is therefore a convenient
way of exciting molecules close to a surface.

7.4 Surface plasmon polaritons

Under certain conditions, electromagnetic surface waves can propagate along
the interface between two media (Raether, 1977). The requirement is a combination of a lossless medium with a positive permittivity with another of
negative real part and positive imaginary part. Although this is an idealized
situation (as all materials exhibit loss) it is close to the practical system of a
dielectric and metal at frequencies below the *plasma frequency*. The modes
are called *surface plasmons* or *surface plasmon polaritons* (SPPs): 'plasmon'
because the optical properties of the metal are consistent with its electrons
behaving as a free electron plasma and 'polariton' to suggest the coupling
of photons with polar excitations in the metal.

Optical excitation of SPPs requires matching of both energy (i.e., optical
frequency) and momentum p. The latter quantity is related to the wavelength

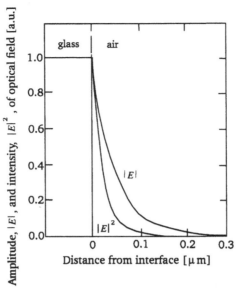

Figure 7.5 Decay of amplitude |E| and intensity |E|² of an evanescent wave at a glass/air interface. (After, Swalen, 1986.)

of the light by *de Broglie's relationship*

$$p = \frac{h}{\lambda} \tag{7.2}$$

where h is Planck's constant. The momentum can also be expressed in terms of the *wavevector k*, where

$$k = \frac{2\pi}{\lambda} = \frac{2\pi p}{h} \tag{7.3}$$

Figure 7.6 contrasts the wavevector versus frequency relationships for light in air (ω/k = velocity of light) and surface plasmons; such curves are called *dispersion curves*. These two curves do not intersect at any frequency. To be able to couple the light to the SPPs, the momentum of the photons at a particular frequency must be increased. This is accomplished by decreasing the velocity of the light, i.e., slowing the photons. As figure 7.6 reveals, the dispersion curve for light in glass now intersects that of the surface plasmons.

Figure 7.7 shows an experimental arrangement for the excitation of surface plasmons. This is known as the *Kretschmann configuration* and exploits evanescent electromagnetic waves. Monochromatic light (*p*-polarized) enters the prism and is refracted onto the base and therefore onto the metal film, where it is reflected. The resulting evanescent field penetrates to the opposite side of this film (i.e., the metal/air interface) if the metal film is

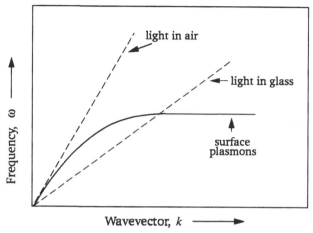

Figure 7.6 Dispersion curves for light in air and glass and for surface plasmons.

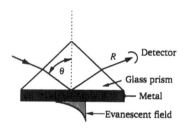

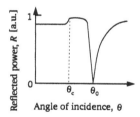

Figure 7.7 Kretschmann configuration for exciting surface plasmons.

sufficiently thin. Changing the angle of incidence of the light alters the component of the wavevector parallel to the prism base. When this component matches the real part of the surface plasmon wavevector (at θ_0), surface plasmons are excited. This process may be detected by a pronounced reduction in the intensity of the reflected beam, R as shown in the bottom diagram of the figure. The phenomenon is referred to as *surface plasmon resonance* (SPR). Gratings may also be used to couple light to SPPs (Raether, 1977).

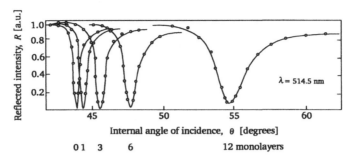

Figure 7.8 Surface plasmon resonance data for silver films coated with different numbers of monolayers of cadmium eicosanoate. Symbols are experimental points; solid lines are theoretical curves. (After Pockrand et al., 1977. Reproduced with permission from Elsevier Science.)

A thin dielectric coating on top of the metal film will shift the surface plasmon dispersion curve to higher momentum. Consequently, the SPR will move to a higher angle. This can be conveniently demonstrated using LB layers. Figure 7.8 shows the experimental data for cadmium eicosanoate monolayers deposited onto silver (Pockrand et al., 1977). The optical constants (refractive index and film thickness) for the organic film can be found from these curves by using a least-squares fit to the exact Fresnel equations (appendix B). Calculations may be performed for both isotropic and uniaxial anisotropic refractive index models. However, the anisotropic calculation does not necessarily provide a better fit to the experimental reflectivity curves than the isotropic calculation.

The shift of the SPR conditions with film thickness can be used as the basis for a high contrast microscope (Knoll, 1991). Here the reflected and scattered plasmonic light is converted by a lens to form an image of the interface on a television camera, for example. Only the areas at resonance appear dark. Variations in film thickness of as little as a few tenths of a nanometre are enough to generate sufficient contrast for an image.

Shifts in the SPR curves may also be predicted if the refractive index of the overlayer is altered. The sensitivity is noteworthy, with claims that changes in refractive index of 10^{-5} may be monitored (Liedberg et al., 1983). If the overcoating on the metal film is absorbing at the wavelength of measurement, then the depth and width of the SPR curve are affected as well as its coupling angle. For example, figure 7.9 shows SPR curves for one and two LB layers of a copper phthalocyanine, at a wavelength of 633 nm (Zhu et al.,1990).

Uses of SPR include sensing. In this type of application changes in the optical properties of an 'active' layer in response to external ambients are detected. The pioneering work in this area was carried out by the Linköping

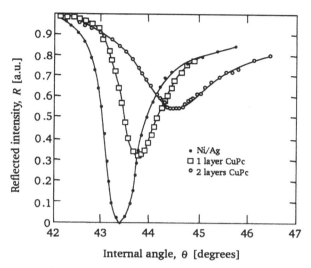

Figure 7.9 Surface plasmon resonance curves (at 633 nm) for LB layers of a copper phthalocyanine deposited onto Ag/Ni. (After Zhu et al., 1990. Reproduced with permission from Elsevier Science.)

group who reported on both gas sensing and biosensing (Liedberg et al., 1983; Nylander et al., 1983). By choosing an angle of incidence close to the plasmon resonance angle, small changes in the resonance conditions (i.e., resonance angle) produced large changes in the reflected intensity that were easily measured. Using a spin-coated active layer consisting of a thin oil film of a silicon-glycol co-polymer, the Linköping workers could detect anaesthetic halothane gas at concentrations ranging from a few per cent to 100 vapour parts per million [v.p.m.], with an estimated lower detection limit of 10 v.p.m. Phthalocyanine LB films have also been used as the gas sensing layer in detectors based on SPR (Lloyd et al., 1988, Zhu et al., 1990). A lower detection limit for nitrogen dioxide was estimated as a few vapour parts per million, which is of the same order as the resistor device discussed in section 6.8.1.

7.5 Optical waveguiding

A further method for studying the optical properties of LB films is based on *optical waveguiding* (Swalen, 1986). EM radiation at optical frequencies can be confined to thin slabs or fibres if the refractive indices of the guiding material and its surroundings are chosen so that *the radiation is always totally internally reflected at the boundaries.* Figure 7.10 shows an example of a planar waveguide formed by a film of refractive index n_f on top of a substrate of refractive index n_s. The topmost layer (refractive index n_0)

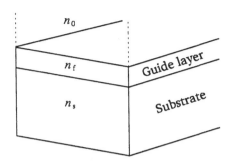

Figure 7.10 A slab planar waveguide.

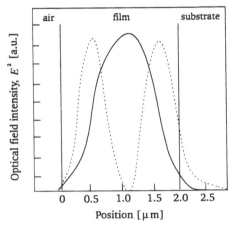

Figure 7.11 The optical field intensity E^2 for two modes of a 2 μm film. The solid line is the m = 0 mode. The m = 1 mode (dashed line) has a minimum in the electric field intensity near the middle of the guide.

is often air and has a much lower refractive index than the other two. Waveguiding will occur in the film providing that its thickness is given by (Wilson and Hawkes, 1983)

$$d \geq \frac{\lambda(m + 0.5)}{2\sqrt{(n_f^2 - n_s^2)}} \tag{7.4}$$

where λ is the wavelength in the guide and m is an integer. Each value of m is associated with a distinct wave pattern or *mode* within the waveguide. Transverse magnetic (TM) and transverse electric (TE) radiation give rise to two types of mode (appendix B); these are called TM_m and TE_m. The cut-off thickness for waveguiding in the asymmetric structure shown in figure 7.10 is usually about one third of the wavelength of the light. Figure 7.11 shows how the optical electric field intensity (E^2) varies with position in a 2 μm thick guiding layer for the TE_0 and TE_1 modes (Swalen, 1986). The fields

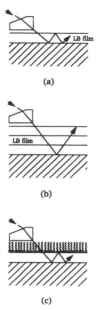

Figure 7.12 Waveguide experiments using LB films: (a) waveguiding in LB film; (b) LB film as part of a laminate waveguide; (c) monolayers probed by evanescent tail of light being guided.

penetrating both the air and the substrate are evanescent in nature; this may be exploited for coupling EM radiation into or out of the guides. Different modes propagate along a waveguide with different velocities, even if they are generated by monochromatic radiation. This phenomenon is called *mode dispersion*.

Figure 7.12 shows waveguiding experiments that may be performed using LB films (Swalen, 1986). The light is coupled into or out of the guide using a prism technique, very similar to the method described in the previous section for coupling to surface plasmons. In figure 7.12(a), the EM radiation is being guided in the LB layer while in figure 7.12(b) the LB layer is part of a laminate structure. Figure 7.12(c) shows a monolayer film deposited on top of a waveguide; here the organic film is probed by the evanescent tail of the waveguiding mode.

As EM energy propagates along a waveguide, its intensity will be reduced. If a beam of intensity I_i is launched into one end of a waveguide and if the intensity at a distance L metres along the guide is I_f, then the *attenuation coefficient* α is given by

$$\alpha = \frac{10\log_{10}(I_f/I_f)}{L} \text{ dB m}^{-1} \qquad (7.5)$$

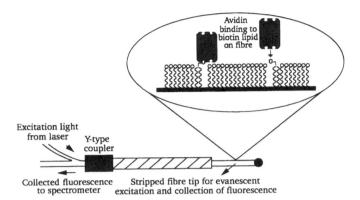

Figure 7.13 Schematic diagram of an evanescent fibre optic sensor used to excite and collect the fluorescence of surface bound avidin to a biotinylated phospholipid. Reprinted with permission from Zhao, S. and Reichert, W. M. (1992) Langmuir, 8, 2785–91. Copyright 1992 American Chemical Society.)

Modern fibre optic cables have α values of the order $1\,\mathrm{dB\,km^{-1}}$. A significant problem with LB film waveguides has been to reduce the attenuation to an acceptable level.

Waveguiding has been observed in passive (e.g., fatty acid) and centrosymmetric LB film systems by several groups (Petty, 1990). There are also reports on successful waveguiding in thick LB films of second-order nonlinear optical materials (see section 7.7 below). Bosshard et al. (1990) have noted that TE and TM modes propagating over more than 20 mm could be excited in thick LB films of a dye material. Attenuation coefficients down to $12\,\mathrm{dB\,cm^{-1}}$ were measured at a wavelength of 633 nm, suggesting that low values, of the order $0.3\,\mathrm{dB\,cm^{-1}}$ near $1.3\,\mu\mathrm{m}$, may be achievable.

Cresswell et al. (1992) have combined the LB deposition technique with a passive polymer dip-coating method to produce a laminated structure (figure 7.12(b)). The thicknesses and refractive indices of the various layers were chosen to give a significant proportion of the guided energy within the LB layer. After fabrication, the waveguide (deposited onto Si/SiO_2) was cleaved to give high quality waveguide faces for optical coupling; the guides were found to support both TM and TE modes.

The evanescent excitation of LB molecules deposited onto an optical fibre waveguide has been used in a sensing device (Zhao and Reichert, 1992). Figure 7.13 shows the experimental set-up. Solutions of fluorescein-labelled avidin (section 4.8.3) were exposed to a series of biotin-doped LB films deposited onto the tip of the fibre. Optical excitation from a laser produced fluorescence from the avidin that had bound to the biotin. The resulting collected fluorescence was then a direct measure of the degree of protein binding.

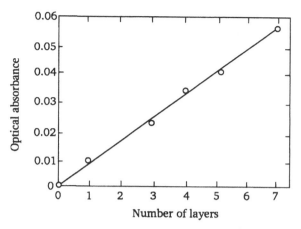

Figure 7.14 Optical absorbance at 332 nm versus number of LB layers for an organotransition metal compound. (After Richardson et al., 1988. Reproduced with permission from Elsevier Science.)

7.6 Dye layers

A straightforward test of the reproducibility of the transfer of a monolayer containing a dye chromophore from the water surface to a solid substrate is to monitor the optical absorption (at a particular wavelength) as a function of the number of LB layers deposited. Figure 7.14 shows the result of such an experiment, using an organotransition metal complex (Richardson et al., 1988). The linear increase in absorption with increasing film thickness agrees with Beer's law (appendix B).

The absorption spectrum of an LB film containing dye molecules may well differ from that measured for a solution of the dye. This, in turn, may not correspond directly with the spectrum of dye molecules in the bulk solid. Although the forces between molecules in crystals are weak and short range (van der Waals' interactions) and the overlap between adjacent molecules in the lattice is small, there is still a substantial difference between the electronic spectra of molecular crystals and free molecules. Some of these differences are caused by interactions between the electronic states of molecules in the vicinity; others are due to crystal lattice properties. Crystal spectra have absorption bands that are broader than the same bands for the molecules in solution. This is because the molecular interactions are affected by thermal vibrations as well as the relative orientation of the molecules. Bands in crystal spectra are often shifted and split when compared to the bands in spectra from the molecules in solution.

Sometimes changes are observed in the absorption spectra of dye LB films over time. These may be the result of physical or chemical changes in the organic film. Figure 7.15 shows an example: the absorption spectra are for a

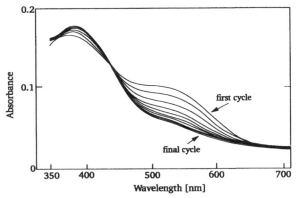

Figure 7.15 Optical absorption spectra for a 78-layer LB film of a mero-cyanine dye: scan rate $1\,nm\,s^{-1}$; cycle time 30 min. (After Neal, 1987.)

78-layer film of a merocyanine dye, measured in air at 30 min intervals (Neal, 1987). The change in the spectrum is due to *protonation* of the dye in air. The invariant point at approximately 442 nm in the series of absorption scans is called an *isosbestic point*. Its presence is consistent with an equilibrium between two different forms of the dye, D (red) and DH$^+$ (yellow), i.e.,

$$D + H^+ \rightleftharpoons DH^+ \tag{7.6}$$

7.6.1   Aggregate formation

The formation of aggregates is quite common in concentrated dye solutions and is often difficult to avoid when the molecules are highly ordered in a built-up Langmuir–Blodgett film. The interactions between dye molecules can be explained by considering energetically delocalized states termed *excitons*. In figure 7.16, E_{excited} and E_{ground} are the unperturbed energy levels of an isolated molecule in dilute solution. When these molecules are brought close together (figure 7.16(a)) the dipoles interact and multiple excitation energy levels are observed. The splitting of energy levels (*Davydov splitting*) is determined by the difference between the interacting transition dipole moments, their relative orientations and the number of interacting molecules. For a simple dimer, the energies are given by

$$\Delta E_{\text{dimer}} = \Delta E_{\text{monomer}} + D \pm \epsilon \tag{7.7}$$

where ΔE represents the transition energies to the excited state of the dimer or monomer, ϵ is the exciton interaction energy and D is a *dispersion energy* term, given by $W' - W$ in the figure; this depends on van der Waals' interactions between the molecules.

Equation (7.7) shows that for a dimer the excited energy state has two possible levels. These can be attributed to the phase relationship between the

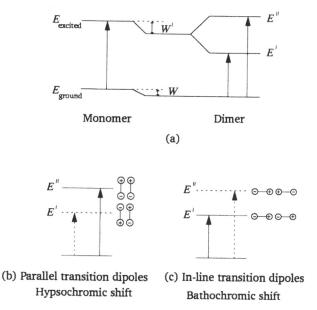

Figure 7.16 Aggregation effects on the energy levels of molecules.

transition dipole moments. If the dipole moments of the two molecules are parallel, as shown in figure 7.16(b), the net transition moment of the dimer is zero in the lower energy state E' and this transition is therefore forbidden. Only transitions to the higher energy state E'' are allowed, leading to a *hypsochromic shift* (*H-band*) which is often observed in LB films. If the transition dipoles are in-line rather than parallel (figure 7.16(c)), transitions to the higher energy state are forbidden and a *bathochromic* shift is observed in the spectrum. Higher aggregates of this type exhibit a large bathochromic shift, with an intense, narrow absorption; these are termed *J-aggregates*, after Jelley (Kuhn et al., 1972). Of course, parallel and in-line are the two extreme forms of molecular aggregation and, in general, the transition dipoles are oriented at some angle to each other, resulting in more complex spectral changes.

Figure 7.17 contrasts the absorption spectra of an amphiphilic hemicyanine dye in LB film form and in a chloroform solution (Neal et al., 1989). The first absorption band is broadened and considerably blue-shifted in the multilayer compared with the solution. This is probably the result of H-aggregates in the LB film. If the monolayer is diluted with a long-chain fatty acid, then the absorption spectrum moves back towards that observed for the solution spectrum (Neal, 1987; Schildkraut et al., 1988).

The effect of J-aggregates is illustrated in figure 7.18(a), which shows the absorption spectrum of a monolayer of an amphiphilic cyanine dye mixed in a 1:1 molar ratio with octadecane (Kuhn et al., 1972). A very narrow

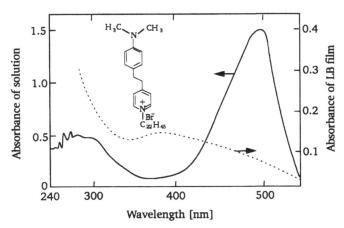

Figure 7.17 Optical absorption spectra for an amphiphilic hemicyanine dye in a chloroform solution (solid line) and in LB film form (dashed line). LB film thickness – 36 layers. The arrows indicate which axis refers to each curve. (After Neal et al., 1989. Reproduced with permission from Elsevier Science.)

absorption band is due to an in-phase relationship between the oscillators corresponding to the dye chromophores. The closest packing of the chromophores is achieved with the brickwork arrangement of dye molecules (figure 7.18(b)). The octadecane chains fit into the cylindrical holes left by the hydrocarbon chains of the dye. Such spectra depend crucially on the monolayer coating conditions and on the nature of the adjacent layer.

7.6.2 Chromophore orientation

The orientation of the chromophores in monolayer or multilayer films containing dyes can be studied by measuring the absorption spectrum for different polarizations and angles of incident radiation. Consider, for example, an amphiphilic azobenzene dye that forms monolayers with an orientation of transition moments (appendix B) normal to the surface if the dye molecules associate to form aggregates. The absorption spectra of a mixed monolayer of such a dye with hexadecane, measured under different conditions, are shown in figure 7.19 (Möbius, 1990). Curve 1 is the absorption spectrum at normal incidence and curve 2 is that obtained using s-polarized light (appendix B) at an angle of incidence of 45°. Both curves show a broad absorption maximum at 470 nm. This can be attributed to dye monomers oriented in the film plane. However, a different spectrum is noted with p-polarized light incident at 45°. This is attributed to aggregates of the dye, oriented with their transition moments normal to the layer plane. Similar studies may be undertaken by monitoring the reflection of polarized light from

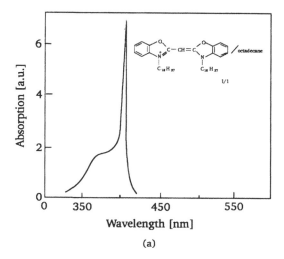

(a)

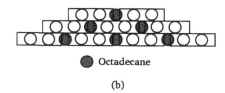

● Octadecane

(b)

*Figure 7.18 (a) Narrow absorption band resulting from J-aggregate forma-
tion in a mixed monolayer of an amphiphilic cyanine dye and octadecane.
(b) The molecules are assumed to be packed in the brickwork arrangement
shown (viewed from above). (After Kuhn, H., Möbius, D. & Bücher, H.
(1972) in* Physical Methods of Chemistry, *Vol. 1, Pt. 3B, eds. A. Weissberger and B. Rossiter. Copyright © John Wiley & Sons, Inc. Reprinted by
permission of John Wiley & Sons, Inc.)*

floating monolayers (Möbius, 1990). For *p*-polarized radiation incident at the
Brewster angle (appendix B), no reflection will occur. A floating monolayer
will perturb this situation. This is exploited in a Brewster angle microscope
(Hönig and Möbius, 1992). Micrographs from such a system have been
shown in chapter 2, figure 2.11.

7.6.3 Energy transfer

Steady state and dynamic fluorescence measurements of LB films are
relatively easy to perform and provide a large amount of structural data
(Möbius, 1990). Energy transfer from monolayers of sensitizer molecules to
a monolayer of acceptor molecules has also been extensively investigated
by Kuhn and coworkers (1972). Figure 7.20 illustrates the type of experiment
that may be undertaken. A sensitizer molecule within a monolayer is

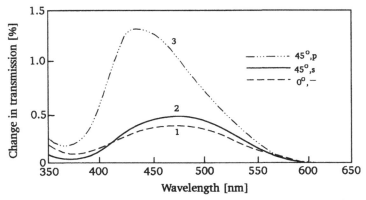

Figure 7.19 Polarized absorption spectra for a mixed monolayer of an amphiphilic azobenzene dye and hexadecane. (After Möbius, 1990. Reproduced with permission from Plenum Press.)

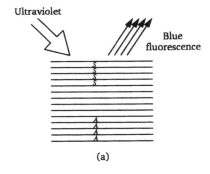

(a)

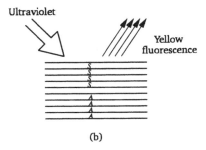

(b)

Figure 7.20 Monolayers of sensitizing dye S and an acceptor molecule A separated by spacer monolayers of a fatty acid. In (a) no interaction is possible. In (b) the spacing is close enough for quantum mechanical tunnelling and energy transfer to occur between A and S molecules. (After Roberts, 1990. Reproduced with permission from Plenum Press.)

represented by S and an acceptor molecule by A. The S molecules absorb in the ultraviolet part of the EM spectrum and fluoresce in the blue, while A absorb in the blue and fluoresce in the yellow. If there is sufficient distance between S and A, the fluorescence of S only appears. However, if the distance between the monolayers of S and A is reduced sufficiently, the fluorescence of S is absorbed by A and yellow fluorescence is observed. The LB technique allows S and A molecules to be at quite precise distances from one another by using monolayers of fatty acids as spacer layers. This simplifies the study of the energy transfer process. The fluorescence intensity I_d for a distance d between sensitizer and acceptor layer is given by (Kuhn et al., 1972)

$$\left(\frac{I_d}{I_\infty}\right) = \left[\left(1 + \left(\frac{d_0}{d}\right)^4\right)\right]^{-1} \tag{7.8}$$

$$d_0 = \alpha\left(\frac{\lambda_S}{n}\right)(q_S F_{AS})^{\frac{1}{4}} \tag{7.9}$$

At distance $d = d_0$, an average of one out of two S molecules loses its excitation energy by energy transfer to acceptor A. According to equation (7.9), d_0 depends on the quantum yield q_S of the fluorescence of the emitter without the acceptor, on the fluorescence absorption F_{AS} of the absorber layer for light of vacuum wavelength λ_S, and on the refractive index n of the medium. The value of the numerical factor α depends on the direction of the transition moments of S and A.

7.6.4 Effect of electric fields on absorption spectra

Changes in the optical properties of organic molecules may be produced by applying large electric fields. Electroabsorption is one technique that has been used to study multilayer films. This is a particular branch of the group of experimental methods known as modulation spectroscopy and involves monitoring small changes in a sample's optical transmission that result from the application of an external field. For molecular materials, these changes are usually the result of the *Stark effect*. The change in potential energy ΔU that arises when a molecule is placed in an electric field may be written

$$\Delta U = -\mu E \cos\theta + \frac{\alpha}{2}E^2 \tag{7.10}$$

where the dipole moment μ makes an angle θ to the applied field and α is the polarizability. The first term on the right-hand side of equation (7.10) represents the linear Stark effect and the second term the quadratic effect. In general, crystals with polar space-groups exhibit a first-order Stark effect. This is usually recognized in electroabsorption spectra as the second derivative of the zero-field absorption curve. Blinov et al. (1983) have reported a study of the linear Stark effect in multilayers of an amphiphilic dye. By varying the deposition conditions, X-, Y- and Z-type multilayers of this material could be fabricated. The reversal of the linear Stark spectra for X- and Z-layers provided direct evidence for the polar nature of these films. The intensity of the signal for the Y-layers was approximately 50 times less than for the X- or Z-films.

Crystal structures having nonpolar spacegroups may still exhibit a second-order Stark effect. This process is recognized in electroabsorption as the first derivative of the zero-field absorption curve and has been noted for anthracene LB films (Roberts et al., 1979).

7.7 Second-order nonlinear optical behaviour

7.7.1 Materials

Organic molecules exhibiting second-order nonlinear optical (NLO) properties (e.g., second-harmonic generation) must be noncentrosymmetric (section B.4). The general approach to the synthesis of NLO LB molecules has been to couple a long hydrocarbon chain onto an asymmetric dye molecule, e.g., an azobenzene or stilbene moiety (Zyss, 1985; Swalen, 1986; Ulman, 1991; Prasad and Williams, 1991; Allen, 1992; Tredgold, 1994).

The second-order nonlinear hyperpolarizability of the individual molecules in an organic solid is measured in terms of the β coefficient (appendix B) (SI units [$C\,m^3\,V^{-2}$]; multiply by 2.7×10^{20} for e.s.u.), while the macroscopic manifestation of the individual molecular contributions is measured as the second-order nonlinear susceptibility $\chi^{(2)}$ (SI units are [$C\,V^{-2}$]; multiply by 2.7×10^{14} for e.s.u.). Other important quantities are the linear electro-optic coefficients r and second-harmonic coefficients d; these are related to $\chi^{(2)}$ by

$$\chi^{(2)}(-\omega; \omega, 0) = \epsilon_0 n_\omega^2 n_0^2 r(-\omega; \omega, 0)/2 \qquad (7.11)$$

and

$$\chi^{(2)}(-2\omega; \omega, 0) = 2\epsilon_0 d \qquad (7.12)$$

n_ω is the appropriate refractive index at the frequency, ω, shown. (The SI units for both the d and r coefficients are [$m\,V^{-1}$].) The frequency arguments in parentheses and to the right of the semicolon represent the applied fields whereas the field resulting from the interaction (which may be regarded as emitted, hence the negative sign) is given to the left of the semicolon. The quantities β, χ, d and r are all tensors and a rigorous formulation of equations (7.11) and (7.12) should include this.

Figure 7.21 compares the second-harmonic d coefficients for various inorganic and organic materials (Petty, 1992). The largest reported

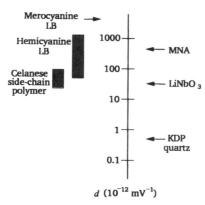

Figure 7.21 Experimental values for the highest second-harmonic d coefficient for various organic and inorganic materials. KDP is potassium dihydrogen phosphate; MNA is 2-methyl-4-nitroaniline. The shaded bars indicate the variation in values for the hemicyanine LB film and the celanese side-chain polymer. (After Swalen et al., 1991; Petty, 1992.)

second-order nonlinear hyperpolarizability is that of the merocyanine dye. Unfortunately, this compound is unstable in air: exposure to the atmosphere results in protonation (figure 7.15), reducing the nonlinear activity. Amphiphilic hemicyanine derivatives have also been extensively investigated (figure 7.17 and chapter 4, figure 4.8). The d coefficient for the most well studied compound, measured with a 1.064 μm incident laser, is about an order of magnitude less than that for the merocyanine dye (Girling et al., 1985). However, no short term chemical reactions have been noted for this compound and a monolayer of it is often used as a standard against which new materials can be rapidly assessed (Petty, 1992). It is interesting that if this compound is diluted in a fatty acid matrix then the second-harmonic generation (SHG) is enhanced. This is a result of the aggregation effects discussed in section 7.6.1. The pure dye is almost completely in the form of H (blue-shifted) aggregates, whereas in a mixed film the dye is mostly monomeric. Diluting the hemicyanine with the fatty acid shifts the main absorption band towards 532 nm, the second-harmonic wavelength of the infrared incident laser. The amount of second-harmonic generation is thereby increased by a resonant effect.

To achieve more stable LB systems for nonlinear optics, polymerizable amphiphiles have been investigated. Chromophores have also been attached to polymeric backbones: examples include polysiloxanes, polyethers and copolymers of styrene-maleic anhydride (Ulman, 1991; Petty, 1992; Tredgold, 1994). Polymers are less tractable than their monomeric counterparts. This has led to the development of some oligomer materials (e.g., those shown in chapter 4, figure 4.15).

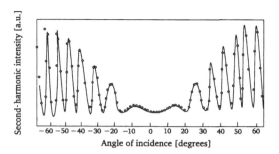

Figure 7.22 Experimental (points) and theoretical (solid curve) data for second-harmonic intensity of an oligomeric LB film as a function of the angle of incidence of the fundamental laser beam. (After Allen, 1992. Reproduced with permission from Research Studies Press.)

7.7.2 Second-harmonic generation

Second-harmonic generation is a simple method of characterizing monolayer and multilayer films. The basic experimental arrangement involves measuring the intensity of the second-harmonic signal generated when the LB film is exposed to a fixed wavelength laser beam (usually an Nd : YAG laser operating at 1.064 µm). The intensity of the second-harmonic signal $I_{2\omega}$, produced by an incident beam of intensity I_ω passing through a material of thickness t, possessing a nonlinear optical coefficient d and a refractive index n is given by (Allen, 1992)

$$I_{2\omega} = 2(\omega t)^2 \left(\frac{\mu_0}{\epsilon_0}\right)^{\frac{3}{2}} \left(\frac{d^2}{n_\omega^2 n_{2\omega}}\right)(I_\omega)^2 \left[\frac{\sin^2(\delta k t/2)}{(\delta k t/2)^2}\right] \qquad (7.13)$$

where δk is a wavevector mismatch term given by $\delta k = k_{2\omega} - 2k_\omega$. The final term on the right-hand side of equation (7.13) is a *sinc* function (sinc$(x) = \sin(x)/x$), which has a maximum value of unity at $x = 0$.

Measurements can be made either at a fixed angle of incidence (using two perpendicular polarizations of the laser beam) or as a function of the angle of incidence (Allen, 1992). In the latter case, the output second-harmonic intensity has the form shown in figure 7.22. The fringing effect is due to interference between the harmonic signals produced from the LB films on the front and back surfaces of the substrate. It is the height and shape of the envelope of these fringes that is of interest. Comparison of the signal obtained with that from a standard sample (e.g., a quartz crystal) allows the d coefficient of the film to be found (Allen et al., 1988). The shape of the envelope, as defined by the position and angular half-width of the peak, can be used to find the tilt angle of the chromophore parts of the LB film with respect to the film normal. From these data it is possible to calculate the β coefficient; typical figures for LB film materials are of the order 10^{-48} C m^3 V^{-2} (Prasad and

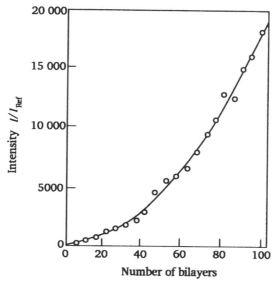

Figure 7.23 Second-harmonic intensity (relative to a hemicyanine standard) versus the number of bilayers of an amphiphilic hemicyanine dye alternated with a zwitterionic 'two-legged' spacer molecule. Points are experimental; the full line corresponds to a quadratic dependence. (After Ashwell, G. J., Dawney, E. J. C., Kuczynski, A. P. & Martin, P. J. (1991) Proc. SPIE Int. Soc. Opt. Eng., 1361, 589–98.)

Williams, 1991; Ulman, 1991; Petty, 1992). Values an order of magnitude greater have been noted but, in most of these cases, the second-harmonic measurements are made in an absorption band of the dye and the β coefficient is increased by a resonant effect.

To exploit the high nonlinear optical coefficients of LB molecules, a means has to be found of building up films of micrometre dimensions (e.g., for waveguiding applications) while preserving the molecular alignment. To have a finite $\chi^{(2)}$, the molecules must be arranged in a noncentrosymmetric manner. This may be achieved by alternating an 'active' material with a 'passive' (e.g., fatty acid) spacer layer; alternating two active materials, chosen so that their β coefficients are additive (e.g., chapter 4, figure 4.15); or exploiting compounds that deposit as X-type or Z-type films. It is possible to observe second-order effects in Y-type layers; this has been attributed to a 'herring-bone' arrangement of the chromophores (Decher et al., 1988).

If the molecular alignment is preserved as the LB film is built up, then a quadratic relationship between the second-harmonic intensity and the film thickness should result (in equation (7.13), the sinc function can be approximated as unity for very thin films). Figure 7.23 shows an example of this, obtained for an amphiphilic hemicyanine dye alternated with a zwitterionic 'two-legged' spacer molecule (Ashwell et al., 1991). Hodge et al. (1994) have reported this quadratic relationship for up to 600 layers (film thickness $\approx$ 1.5 μm) of alternating active/passive polymer films. Unfortunately, there are many other instances of multilayer systems that do not show such quadratic behaviour. The molecules in LB films tend to re-arrange, either during transfer from the water surface or later, to minimize their free energy (section 3.3).

7.7.3 Pockels' effect

The SPR technique discussed in section 7.4 can be adapted for measurement of the electrooptic coefficient of an LB film. An electric field E is applied to the organic film as the angle of incidence of the p-polarized light is changed. A small change in permittivity, due to the Pockels' effect, is produced. The second-order nonlinear susceptibility is given by (Cresswell, 1992)

$$\chi^{(2)}(-\omega; \omega, 0) = \frac{\Delta\epsilon' + j\Delta\epsilon''}{2E} \qquad (7.14)$$

SPR and Pockels' data are shown in figure 7.24 for monolayers of hemicyanine and nitrostilbene dyes mixed with a fatty acid (Petty and Cresswell, 1991). In contrast to second-harmonic measurements, the Pockels' experiment gives both the magnitude and *the sign* of the molecular hyperpolarizability. Comparison of figure 7.24(a) and (b) reveal that the β coefficient for the hemicyanine layer is in the opposite sense to that of the nitrostilbene layer (i.e., the differential reflectivity curves for the two dyes are inverted). The magnitudes of the electrooptic r coefficients for the two materials are approximately: 40 pm V^{-1} (hemicyanine) and 8 pm V^{-1} (nitrostilbene). High values have also been found for other monolayer films, such as the oligomers shown in chapter 4, figure 4.15 ($\chi^{(2)} = 10$–20 pm V^{-1}). However, as noted above, the nonlinearities are not always preserved as the film thickness is increased (Cresswell et al., 1994).

7.7.4 Phase-matching

Phase-matching refers to the degree of coherence between the nonlinear optical polarization and the generated light. Equation (7.13) shows that the second-harmonic intensity is only large when $\delta k t = 0$. When the condition $\delta k = 0$ is met, the interaction is phase-matched. Since $k = \omega n / c$

$$\delta k = \left(\frac{2\omega}{c}\right)(n_{2\omega} - n_\omega) \qquad (7.15)$$

The requirement for phase-matching is that the refractive index of the medium is the same at the fundamental and harmonic frequencies. Because all materials are dispersive, it is usual that $n_{2\omega} > n_\omega$. In birefringent media, phase-matching

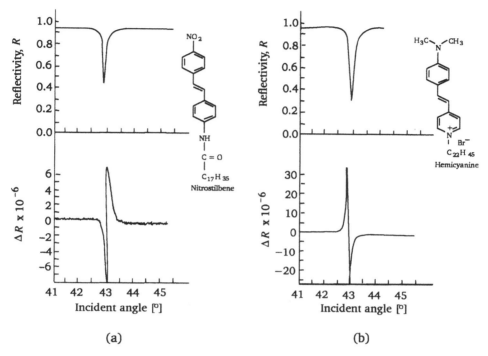

Figure 7.24 *Angular dependence of the reflectivity* (R) *and the differential reflectivity* (ΔR) *for monolayer films of 22-tricosenoic acid mixed with (a) a nitrostilbene compound and (b) a hemicyanine compound. The experimental data and theoretical fits are indistinguishable on the scales shown. (After Petty and Cresswell, 1991. Reproduced with permission from World Scientific Publishing.)*

can be achieved by allowing the fundamental and harmonic waves to have different polarizations. If the birefringence is large enough it may be possible to find some angle at which the extraordinary refractive index at the fundamental is equal to the ordinary index at the second-harmonic frequency (or vice versa).

Phase-matching can also be accomplished in waveguide structures (Bosshard et al., 1991; Prasad and Williams, 1991; Ulman, 1991; Fujiwara et al., 1992; Kowalczyk et al., 1993; Küpfer et al., 1993). There are three main approaches: (i) phase-matching by conversion of a guided fundamental mode into a guided second-harmonic mode using mode dispersion (section 7.5); (ii) modulation of either the refractive index of the waveguide structure or the nonlinear susceptibility; or (iii) conversion of a guided fundamental mode into a second-harmonic radiation mode using *Cerenkov* radiation.

Using mode dispersion, the waveguide thickness has to be perfectly adjusted for phase-matching. An additional requirement is that the *overlap*

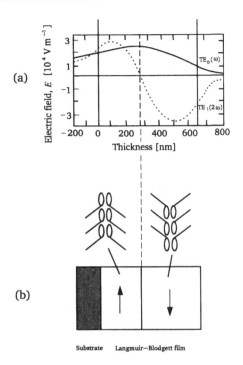

Figure 7.25 (a) Electric field E distribution in film thickness direction (perpendicular to substrate) for a fundamental TE_0 and a second-harmonic TE_1 mode in a substrate/LB film/air waveguide. The full vertical lines represent the substrate/LB film and LB film/air interfaces. (b) Configuration with nonlinear susceptibility inversion in LB layer showing orientation of molecules. (After Küpfer et al., 1993.)

integral S of the guide must be optimized (Ulman, 1991). This is given by (Bosshard et al., 1991)

$$S = t^{1/2} \int_0^t d(E_n^\omega)^2 E_m^{2\omega} \, dz \qquad (7.16)$$

where m, n are the mode numbers, t is the waveguide thickness, d is the second-harmonic coefficient and E is the electric field in the film thickness direction. The optimization of the overlap integral by a proper choice of interacting guided modes is the most important requirement for efficient frequency doubling in waveguides.

Figure 7.25(a) shows an example of the electric field distribution in a waveguide in which a TE_0 fundamental is phase-matched to a TE_1 second-harmonic mode. Because the product of the fields changes sign across the waveguide, the value of S is considerably reduced. However, if the sign of the nonlinear optical susceptibility is inverted at a point where the electric field of the

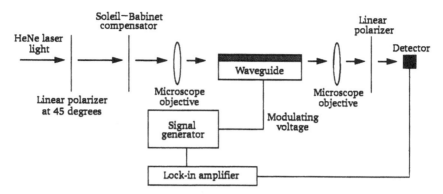

Figure 7.26 Experimental arrangement to demonstrate electrooptic modulation. (After Cresswell et al., 1992. Reproduced with permission from Elsevier Science.)

second-harmonic has a node in the thickness direction (shown by the vertical dashed line in figure 7.25(a)), then the value of S can be increased. Figure 7.25(b) shows how this can be accomplished (Küpfer et al., 1993). This is an example of approach (ii) given above. The waveguide exploits the herring-bone arrangement of molecules, noted in section 7.7.2, and the resulting second-order nonlinear susceptibility is therefore parallel to the substrate. The susceptibility inversion is achieved by simply inverting the substrate at an appropriate point during the LB deposition process.

The Cerenkov technique puts fewer constraints on the design of wave-guides. Radiation from the waveguide exits in a cone in analogy to Cerenkov radiation from charged particles (Feynman et al., 1963). In Cerenkov phase-matching the second-harmonic light propagates as a radiation mode, while the fundamental is guided. This has now been demonstrated in LB film wave-guides (Bosshard et al., 1991; Fujiwara et al., 1992).

7.7.5 Electrooptic modulation

The second-order NLO properties of LB films may be used ito demonstrate electrooptic modulation. Figure 7.26 shows an experimental set-up based on a waveguiding geometry (Cresswell et al., 1992). Light of equal intensities is launched into TM and TE modes and their relative phase adjusted by a *Soleil–Babinet compensator* to give circular polarization at the waveguide output. An a.c. electric field is then applied and the change in transmitted intensity monitored by a lock-in amplifier. Any intensity change is due to modulation of the relative phase of the two polarizations.

Measurements of the electrooptic effect can also be made using *Fabry–Perot*. The multiple pass effect of these devices (known as *etalons*) serves to increase the effective interaction length of an optical beam within the material

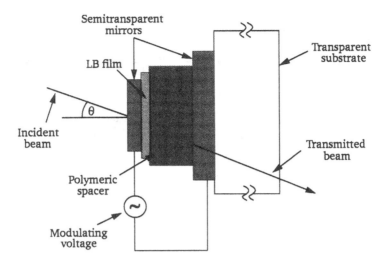

Figure 7.27 Transmission Fabry–Perot etalon. (After Eldering et al., 1989. Reproduced with permission from Elsevier Science.)

and the interferometric nature of the cavity converts phase modulation to amplitude modulation. Eldering et al. (1989) have described the construction and evaluation of such a structure, shown in figure 7.27, in which LB films were incorporated into the spacer layer of the etalon. The angular dependence of the observed transmissivity modulation in the device was used to prove that the electrooptic effect in the LB multilayer was the physical origin of the electrically induced changes in the transmission. A similar device, based on polymer LB layers, has been reported by Cresswell et al. (1995).

7.8 Third-order nonlinear optical effects

Third-harmonic generation does not require a centrosymmetric structure and can be observed in any LB film. The third-order nonlinear susceptibility, $\chi^{(3)}$, is relatively large in polydiacetylenes (chapter 4, table 4.3) and there have been many investigations of this material (Petty, 1990; Prasad and Williams, 1991; Ulman, 1991). However, third-order NLO processes are much less efficient than second-order ones. Consequently, the absolute magnitudes of these effects are quite small.

References

Allen, S. (1992) Nonlinear optics, in *Molecular Electronics*, ed. G. J. Ashwell, pp. 207–65, Research Studies Press, Taunton

Allen, S., McLean, T. D., Gordon, P. F., Bothwell, B. D., Robin, P. and Ledoux, I. (1988) Properties of polyene Langmuir–Blodgett films, *Proc. SPIE Int. Soc. Opt. Eng.*, **971**, 206–15

Allen, S., Ryan, T. G., Hutchings, M. G., Ferguson, I., Swart, R., Froggart, E. S., Burgess, A., Eaglesham, A., Cresswell, J. and Petty, M. C. (1993) Characterization of nonlinear optical Langmuir–Blodgett oligomers, *Proc. OMNO 92 Conference*, eds. G. J. Ashwell and D. Bloor, pp. 50–60, Royal Society of Chemistry, London

Ashwell, G. J., Dawnay, E. J. C., Kuczynski, A. P. and Martin, P. J. (1991) The highest observed second-harmonic intensity from a multilayered Langmuir–Blodgett film structure, *Proc. SPIE Int. Soc. Opt. Eng.*, **1361**, 589–98

Barnes, W. L. and Sambles, J. R. (1987) Surface-pressure effects on Langmuir–Blodgett multilayers of 22-tricosenoic acid, *Surf. Sci.*, **187**, 144–52

Blinov, L. M., Davydova, N. N., Lazarev, V. V. and Yudin, S. G. (1983) Spontaneous polarization of Langmuir multimolecular films, *Sov. Phys. Solid State* (Engl. Trans.), **24**, 1523–5

Blodgett, K. B. (1939) Use of interference to extinguish reflection of light from glass, *Phys. Rev.*, **55**, 391–404

Bosshard, C., Küpfer, M., Günter, P., Pasquier, C., Zahir, S. and Seifert, M. (1990) Optical waveguiding and nonlinear optics in high quality 2-docosylamino-5-nitropyridine Langmuir–Blodgett films, *Appl. Phys. Lett.*, **56**, 1204–6

Bosshard, C., Flörscheimer, M., Küpfer, M. and Günter, P. (1991) Cerenkov-type phase-matched second-harmonic generation in DCANP Langmuir–Blodgett film waveguides, *Opt. Commun.*, **85**, 247–53

Cresswell, J. P. (1992) Waveguiding in electrooptic Langmuir–Blodgett films, PhD thesis, University of Durham, UK

Cresswell, J. P., Cross, G. H., Bloor, D., Feast, W. J. and Petty, M. C. (1992) Electrooptic modulation in polymer/Langmuir–Blodgett film waveguides, *Thin Solid Films*, **210/211**, 216–18

Cresswell, J. P., Petty, M. C., Shearman, J. E., Allen, S., Ryan, T. G. and Ferguson, I. (1994) Electrooptic properties of some oligomeric Langmuir–Blodgett films, *Thin Solid Films*, **244**, 1067–72

Cresswell, J. P., Petty, M. C., Wang, C. H., Wherrett, B., Ali-Adib, Z. and Hodge, P. (1995) An electrooptic Fabry–Perot through-plane-modulator based on a novel Langmuir–Blodgett film, *Opt. Commun.*, **115**, 271–5

Decher, G., Tieke, B., Bosshard, C. and Günter, P. (1988) Optical second-harmonic generation in Langmuir–Blodgett films of 2-docosylamino-5-nitropyridine, *J. Chem. Soc., Chem. Commun.*, 933–4

den Engelsen, D. (1971) Ellipsometry of anisotropic films, *J. Opt. Soc. Amer.*, **61**, 1460–6

Drexhage, K. H. (1966) Habilitahon-Schrift, University of Marburg, Germany

Eldering, C. A., Kowel, S. T., Knoesen, A., Anderson, B. L. and Higgins, B. G. (1989) Characterization of modulated spin-coated and Langmuir–Blodgett thin film etalons, *Thin Solid Films*, **179**, 535–42

Feynman, R. P., Leighton, R. B. and Sands, M. (1963) *The Feynman Lectures on Physics*, Vol. 1, chapter 51, Addison-Wesley, Reading, Massachusetts

Fleck, M. (1966) Dissertation, University of Marburg, Germany

Forster, H. (1966) Diplomarbeit, University of Marburg, Germany

Fujiwara, I., Asai, N. and Howarth, V. (1992) Preparation and characterization of a novel Langmuir–Blodgett film optical waveguide with a nonlinear susceptibility inversion, *Thin Solid Films*, **221**, 285–91

Girling, I. R., Cade, N. A., Kolinsky, P. V., Earls, J. D., Cross, G. H. and Peterson, I. R. (1985) Observation of second-harmonic generation from Langmuir–Blodgett multilayers of a hemicyanine dye, *Thin Solid Films*, **132**, 101–12

Hodge, P., Ali-Adib, Z., West, D. and King, T. A. (1994) Efficient second-harmonic generation from all-polymeric Langmuir–Blodgett 'AB' films containing up to 600 layers, *Thin Solid Films*, **244**, 1007–11

Hönig, D. and Möbius, D. (1992) Reflectometry at the Brewster angle and Brewster angle microscopy at the air–water interface, *Thin Solid Films*, **210/211**, 64–8

Knoll, W. (1991) Optical characterization of organic thin films and interfaces with evanescent waves, *MRS Bulletin*, July 1991, 29–39

Kowalczyk, T. C., Singer, K. D. and Cahill, P. A. (1993) Anomalous dispersion enhanced Cerenkov phase-matching, in *Nonlinear Optical Properties of Organic Materials VI*, ed. G. R. Möhlmann, pp. 332–43, SPIE Proc. 2025, Bellington, Washington, USA

Kuhn, H., Möbius, D. and Bücher, H. (1972) Spectroscopy of monolayer assemblies, in *Techniques of Chemistry*, Vol. 1, Pt. 3B, eds. A. Weissberger and B. Rossiter, pp. 577–702, Wiley, New York

Küpfer, M., Flörsheimer, M., Bosshard, C. and Günter, P. (1993) Phase-matched second-harmonic generation in $\chi^{(2)}$-inverted Langmuir–Blodgett waveguide structures, *Electron. Lett.*, **29**, 2033–4

Liedberg, B., Nylander, C. and Lundström, I. (1983) Surface plasmon resonance for gas detection and biosensing, *Sensors and Actuators*, **4**, 299–304

Lloyd, J. P., Pearson, C. and Petty, M. C. (1988) Surface plasmon resonance studies of gas effects in phthalocyanine Langmuir–Blodgett films, *Thin Solid Films*, **160**, 431–43

Möbius, D. (1990) Spectroscopy of complex molecules, in *Langmuir–Blodgett Films*, ed. G. G. Roberts, pp. 223–72, Plenum Press, New York

Neal, D. B. (1987) Langmuir–Blodgett films for nonlinear optics, PhD thesis, University of Durham, UK

Neal, D. B., Kalita, N., Pearson, C., Petty, M. C., Lloyd, J. P., Roberts, G. G., Ahmad, M. M. and Feast, W. J. (1989) Multilayer assemblies for nonlinear optics, *Synthetic Metals*, **28**, D711–D719

Nylander, C., Liedberg, B. and Lind, T. (1983) Gas detection by means of surface plasmon resonance, *Sensors and Actuators*, **3**, 79–88

Petty, M. C. (1990) Characterization and properties, in *Langmuir–Blodgett Films*, ed. G. G. Roberts, pp. 133–221 Plenum Press, New York

Petty, M. C. (1992) Possible applications for Langmuir–Blodgett films, *Thin Solid Films*, **210/211**, 417–26

Petty, M. C. and Cresswell, J. P. (1991) Deposition and characterization of multilayer films for nonlinear optics, in *Materials for Photonic Devices*, eds. A. D' Andrea, A. Lapiccirella, G. Marletta and A. Viticoli, pp. 259–69, World Scientific Publishing, Singapore

Pitt, C. W. and Walpita, L. M. (1980) Lightguiding in Langmuir–Blodgett films, *Thin Solid Films*, **68**, 101–27

Pockrand, I., Swalen, J. D., Gordon II, J. D. and Philpott, M. R. (1977) Surface plasmon spectroscopy of organic monolayer assemblies, *Surf. Sci.*, **74**, 237–44

Prasad, P. N. and Williams, D. J. (1991) *Introduction to Nonlinear Optical Effects in Molecules and Polymers*, Wiley-Interscience, New York

Raether, H. (1977) Surface plasma oscillations and their applications, *Phys. Thin Films*, **9**, 145–261

Richardson, T., Roberts, G. G., Polywka, M. E. C. and Davies, S. G. (1988) Preparation and characterization of organotransition metal Langmuir–Blodgett films, *Thin Solid Films*, **160**, 231–9

Roberts, G. G. (1990) Potential applications of Langmuir–Blodgett films, in *Langmuir–Blodgett Films*, ed. G. G. Roberts, pp. 317–411, Plenum Press, New York

Roberts, G. G., McGinnity, T. M., Barlow, W. A. and Vincett, P. S. (1979) Electroluminescence, photoluminescence, and electroabsorption of a lightly substituted anthracene Langmuir film, *Solid State Commun.*, **32**, 683–6

Schildkraut, J. S., Penner, T. L., Willard, C. S. and Ulman, A. (1988) Absorption and second-harmonic generation of monomer and aggregate hemicyanine dye in Langmuir–Blodgett films, *Optics Letts.*, **13**, 134–6

Swalen, J. D. (1986) Optical properties of Langmuir–Blodgett films, *J. Molecular Electronics*, **2**, 155–81

Swalen, J. D., Rieckhoff, K. E. and Tacke, M. (1978) Optical properties of arachidate monolayers by integrated optical techniques, *Opt. Commun.*, **24**, 146–8

Swalen, J. D., Bjorklund, G. C., Fleming, W., Herminghaus, S., Jungbauer, D., Jurich, M., Moerner, W. E., Reck, B., Smith, B. A., Twieg, R., Willson, C. G. and Zentel, R. (1991) in *Organic Molecules for Nonlinear Optics and Photonics*, eds. J. Messier, F. Kajzar and P. Prasad, pp. 433–45, Kluwer, Dordrecht

Tomar, M. S. (1974) Skeletonized films and measurement of their optical constants, *J. Phys. Chem.*, **78**, 947–50

Tredgold, R. H. (1994) *Order in Thin Organic Films*, Cambridge University Press, Cambridge

Ulman, A. (1991) *Ultathin Organic Films*, Academic Press, San Diego

Wilson, J. and Hawkes, J. F. B. (1983) *Optoelectronics: An Introduction*, Prentice-Hall, New Jersey

Zhao, S. and Reichert, W. M. (1992) Influence of biotin lipid surface density and accessibility on avidin binding to the tip of an optical fibre sensor, *Langmuir*, **8**, 2785–91

Zhu, D. G., Petty, M. C. and Harris, M. (1990) An optical sensor for nitrogen dioxide based on a copper phthalocyanine Langmuir–Blodgett film, *Sensors and Actuators B*, **2**, 265–9

Zyss, J. (1985) Nonlinear organic materials for integrated optics: a review, *J. Molec. Electr.*, **1**, 25–45

Appendix A

Electronic energy levels in organic solids

A.1 Bonding in organics

Carbon has an atomic number of six and a valency of four. Its electron configuration is $1s^2$, $2s^2$, $2p^2$, i.e., the inner s shell is filled and the four electrons available for bonding are distributed two in s orbitals and two in p orbitals. The s orbital is spherically symmetrical, as shown in figure A.1(a), and can form a bond in any direction. In contrast, the p orbitals, figure A.1(b), are directed along mutually orthogonal axes and will tend to form bonds in these directions. When two or more of the valence electrons of carbon are involved in bonding with other atoms, the bonding can be explained by the construction of *hybrid orbitals* by mathematically combining the 2s and 2p orbitals. In the simplest case, the carbon 2s orbital hybridizes with a single p orbital. Two *sp hybrids* result by taking the sum and difference of the two orbitals, as shown in figure A.2, and two p orbitals remain. The sp orbitals are constructed from equal amounts of s and p orbitals; they are linear and $180°$ apart.

Other combinations of orbitals lead to different hybrids. For example, consider three groups bonded to a central carbon atom. From the 2s orbital and two p orbitals (e.g., a p_x and a p_y), three equivalent sp^2 *hybrids* may be constructed. Each orbital is 33.3% s and 66.7% p. The three hybrids lie in the xy plane (the same plane defined by the two p orbitals), directed $120°$ from each other, and the remaining p orbital is perpendicular to the sp^2 plane. Four sp^3 *hybrids* may be derived from an s orbital and three p orbitals. These are directed to the corners of a tetrahedron with an angle between the bonds of $109.5°$; each orbital is 25%s and 75%p.

197

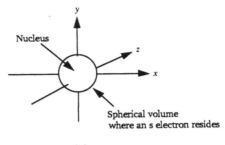

(a)

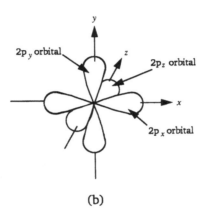

(b)

Figure A.1 (a) s atomic orbital. (b) Mutually orthogonal 2p orbitals.

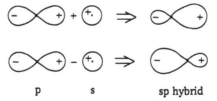

Figure A.2 Mathematical combination of s and p orbitals to give two sp hybrid orbitals.

It is important to note that not all orbitals on an atom are necessarily equivalent when the molecule lacks symmetry, e.g., it is possible to have a hybridized orbital that is 23%s and 77%p. Thus, sp, sp^2 and sp^3 hybrids must be considered as limiting cases only. Electrons in s orbitals have a lower energy than electrons in p orbitals. Therefore, bonds with more s character tend to be stronger.

Long-chain organic compounds, such as the fatty acids, involve C–H bonds that are all approximately $C(sp^3)$–$H(s)$ with a bond length of 0.110 nm and a strength of 4.26 eV (98×10^3 kcal kmole^{-1}). All the C–C

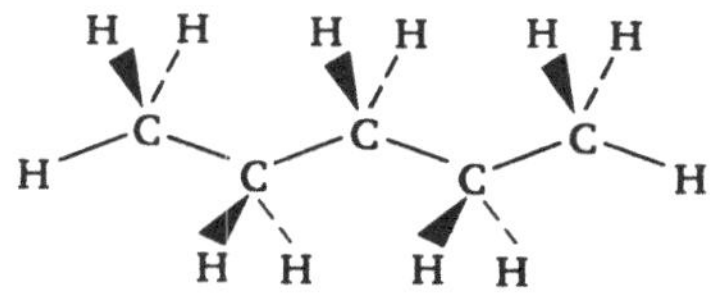

Figure A.3 Covalent bonding in pentane, $CH_3(CH_2)_3CH_3$.

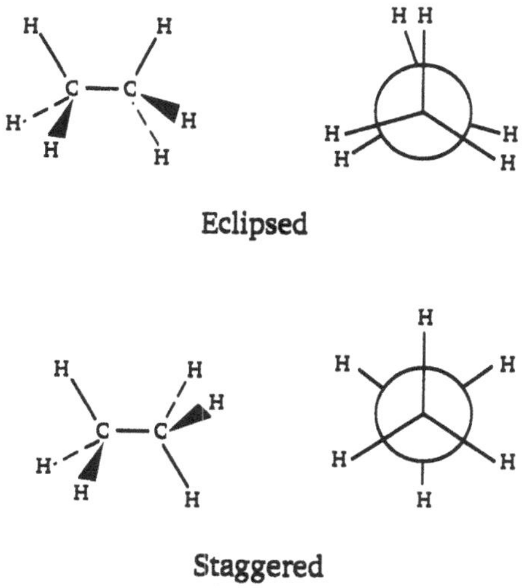

Eclipsed

Staggered

Figure A.4 Eclipsed and staggered conformations of ethane. The Newman projections are shown on the right.

bonds are approximately $C(sp^3)$–$C(sp^3)$ with a length of 0.154 nm and a strength of 3.83 eV (88×10^3 kcal kmole^{-1}). The tetrahedral geometry of the sp^3 orbitals leads to an alkyl chain with a zig-zag conformation, as illustrated for pentane, C_5H_{12} a simple *alkane*, in figure A.3. (Pentane is an example of a straight-chain or unbranched alkane, formally an *n-alkane*.) In this representation, a dashed bond projects away from the viewer, a heavy wedge bond projects towards the viewer, and a normal bond lies in the plane of the page.

Figure A.3 shows the most stable structure of pentane. Rotation about one or more of the single bonds in this molecule leads to different *conformational isomers* or *conformations*. Unlike other kinds of isomers, these cannot usually be isolated as they interconvert too rapidly. The conformation in figure A.3 is called *staggered* and has all the carbon–hydrogen bonds as far away from each other as possible. The least stable conformation is called *eclipsed* and has the carbon–hydrogen bonds as close as possible. Figure A.4 gives a way of representing both the eclipsed and staggered conformations of ethane

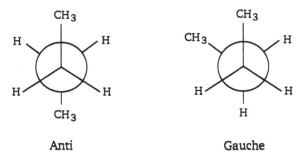

Anti   Gauche

Figure A.5 Anti and gauche conformations of alkanes.

(C_2H_6). These diagrams are called *Newman projections*. The C−C bond is viewed end-on; the nearest carbon is represented by a point and the three groups attached to the carbon radiate as three lines from this point. The farthest carbon is represented by a circle with its bonds radiating from the edge of a circle.

For longer chain alkanes, the conformational situation becomes complex. There are different staggered and eclipsed conformations, not all of equal energy. The lowest energy conformation is the one in which the two large methyl groups are as far apart as possible − 180°; this is the *anti* conformation. A higher energy *gauche* conformation exists when the methyl groups are 60° apart. Newman projections are shown for both these arrangements in figure A.5. Note that there are no eclipsing interactions for the gauche conformation; the increased energy simply arises because the large methyl groups are close together.

The bonding for hydrocarbons with carbon–carbon double bonds (*alkenes*) and triple bonds (*alkynes*) involve sp^2 and sp hybrids. In ethylene ($CH_2=CH_2$), two sp^2 hybrids on each carbon bond with the hydrogens. A third sp^2 hybrid on each carbon forms a $C(sp^2)-C(sp^2)$ single bond, leaving a p orbital 'left over' on each carbon. This orbital lies perpendicular to the plane of the six atoms. The two p orbitals are parallel to each other and have regions of overlap above and below the molecular plane. This type of bond in which there are two sideways bonding regions above and below a nodal plane is called a π-bond. In contrast, a bond formed by the head-on overlap of the two carbon sp^2 orbitals is called a σ-bond. σ- and π-bonds are illustrated in figure A.6(a). The C=C double bond distance in ethylene is 0.133 nm, less than the value of 0.154 nm given above for a C−C single bond, while the C−H bond is 0.108 nm long. Strengths of the C=C and C−H bonds in ethylene are 6.61 eV (152×10^3 kcal kmole^{-1}) and 4.48 eV (103×10^3 kcal kmole^{-1}), respectively.

The noncylindrically symmetric electron density about the C=C bond axis in ethylene (figure A.6(b)) and related compounds leads to a barrier to rotation

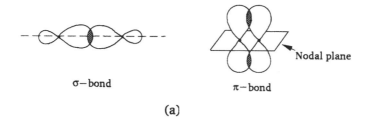

σ−bond   π−bond

(a)

π−bond   p−orbital

H   H

H   H

C(sp²)−H(ls)   C(sp²)−C(sp²) σ−bond

(b)

Figure A.6 (a) σ- and π-bonds; (b) σ−π bonding in ethylene.

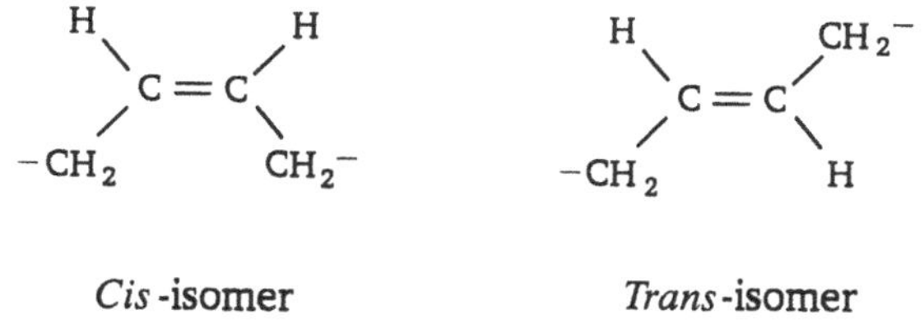

Cis-isomer   Trans-isomer

*Figure A.7 The cis- and trans-isomers for a long-chain hydrocarbon con-
taining a C=C double bond.*

about this axis. Therefore, two isomers exist. As these are not easily
interconverted they are called *configurational isomers,* to distinguish them
from the conformational isomers discussed above, and are known by the prefixes *cis-* (from the Latin for 'on this side') and *trans-* ('across'). Figure A.7
illustrates the *cis-* and *trans-* forms for a long-chain compound containing a
C=C double bond.

In acetylene (HC≡CH), which contains a C≡C triple bond, the C−H bonds
are shorter than in ethylene (0.106 nm) and are C(sp)−H(s) in character. The
bond strength is 5.43 eV (125 kcal kmole^{-1}). The molecule has a linear structure with a C≡C bond of strength 8.7 eV (200×10^3 kcal kmole^{-1}) and length

Table A.1 *Comparison of carbon–carbon and carbon–hydrogen bonds in ethane, ethylene and acetylene.*

Molecule	Bond	Bond strength [eV]	Bond length [nm]
ethane, CH_3CH_3	$C(sp^3)–C(sp^3)$	3.83	0.154
	$C(sp^3)–H(1s)$	4.26	0.110
ethylene, $H_2C{=}CH_2$	$C(sp^2){=}C(sp^2)$	6.61	0.133
	$C(sp^2)–H(1s)$	4.48	0.108
acetylene, $HC{\equiv}CH$	$C(sp){\equiv}C(sp)$	8.70	0.120
	$C(sp)–H(1s)$	5.43	0.106

$H(1s)–C(sp)\ C(sp)–C(sp)\ C(sp)–H(1s)$

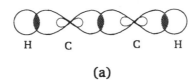

(a)

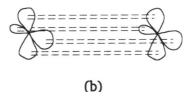

(b)

Figure A.8 *Bonding in acetylene. (a) σ-bonding. (b) π-bonding. The electrons in the two orthogonal p orbitals form a cylindrically symmetric torus or doughnut-like electron distribution.*

0.120 nm, the shortest carbon–carbon bond distance known. A σ-bond is formed from the head-on overlap of the carbon sp hybrid orbitals. The two orthogonal p orbitals give rise to cylindrically symmetrical π-bonds, as shown in figure A.8.

Table A.1 compares the carbon–carbon and carbon–hydrogen bond characteristics for ethane, ethylene and acetylene. In unsaturated alkyl chains, both $-C{=}$ and $-C{\equiv}$ bonds are associated with small, but finite, dipole moments (section 2.2). This is simply because of the different amounts of s character of the carbon orbitals making up the bond. Symmetrically substituted alkenes and alkynes, of course, possess no overall moment.

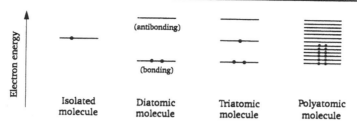

Figure A.9 Splitting of energy levels resulting from overlap of electron orbitals of adjacent atoms. (• – electrons.)

A.2   Band theory

The electrical and optical properties of any solid are determined largely by their electronic structure. The electron distribution may be described by the charge and spatial distribution of the electron orbitals. Whenever two identical atoms are brought into proximity, orbital overlap can occur and each energy level is split into two new levels, with one above and one below the original level.

Consider the formation of a hydrogen molecule from two separated atoms. When the two atoms come together so that their 1s electron orbitals overlap, two new σ-electronic orbitals are formed around the atoms, symmetric with respect to the interatomic axis. In one orbital, the *bonding orbital*, the electron has a lower energy than in the isolated atom orbital and in the other, the *antibonding orbital*, an electron has a higher energy, figure A.9.

In an extended solid many atoms can interact and many similar splittings of energy levels occur. For a solid containing approximately 10^{26} (Avogadro's number) atoms, each energy level splits, but the energies between these split levels are very small and continuous ranges or *bands* of energy are formed, as shown in figure A.9. Two such important bands are the *valence band* and the *conduction band*, analogous to the bonding and antibonding levels of the above two-atom model. The *energy gap*, or *band gap*, between them is a forbidden zone for electrons. Electrical conduction takes place by electrons moving under the influence of an applied electric field in the conduction band and/or *holes* moving in the valence band. Holes are really vacancies in a band. For convenience they may be regarded as positively charged carriers.

Figure A.10 depicts various energy band structures for solids. In a metal, electrons in the highest filled energy level can readily be excited (by an applied electric field) to an unoccupied level since the energy separation is very small. An insulator only has completely filled bands and completely empty ones. In a full band there can be no net flow of electronic charge under an external electric field. The reason for this is that for every wave state in which an electron is travelling in one direction, there is another in

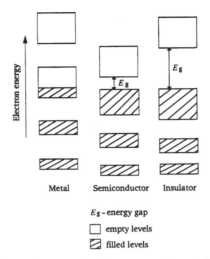

Figure A.10 Electronic energy band structure of a metal, a semiconductor and an insulator.

which the electron is travelling in the opposite direction and there are no spare states. The energy gap in an insulator is large (several electronvolts), preventing an electron being promoted from the top of the highest (in energy) filled band, the valence band, to the bottom of the lowest empty band, the conduction band. For a semiconductor, the band structure resembles that of the insulator, except that the energy gap is much smaller (0.1–2 eV). Charge carriers can be introduced into the valence and conduction bands by departures from stoichiometry, by the addition of impurities, called *doping*, or by the excitation (optically or thermally) of electrons across the band gap.

An important feature of the band system is that the electrons are *delocalized* or spread over the lattice. The strength of the interaction between the overlapping orbitals determines the extent of delocalization that is possible for a given system. The greater the degree of electron delocalization, the larger the width of the bands (in energy terms) and the higher the mobility of the carriers within the band. For many polymeric organic materials, the molecular orbitals responsible for bonding the carbon atoms of the chain together are sp^3 hybridized σ-orbitals that do not give rise to extensive overlapping. The resulting band gap is large, as the electrons involved in the bonding are strongly localized on the carbon atoms and cannot contribute to the conduction process. This is why polyethylene $(CH_2)_n$, for example, is electrically insulating, figure A.11(a).

A significant increase in the degree of electron delocalization may be found in polymers that contain double and triple carbon–carbon bonds. If each

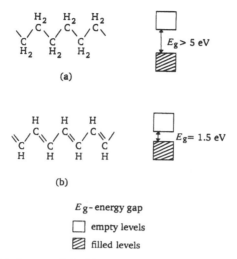

Figure A.11 (a) Saturated backbone structure of polyethylene leads to electron localization and an electrical insulator. (b) Bond alternation in trans-polyacetylene results in a semiconductor band structure.

carbon atom in the chain is attached to its neighbouring carbon atoms and to only one hydrogen, the spare electron in a p_z-orbital of each carbon atom overlaps with those of carbon atoms on either side forming delocalized molecular orbitals of π symmetry, figure A.6(b). For a simple lattice of length $L = Na$, where N is the total number of atoms and a is the spacing between them, it can be shown that the total number of electron states in the lowest energy band is exactly equal to N. This result is true for every energy band in the system and applies to three-dimensional lattices. Allowing for two spin orientations of an electron, the *Pauli exclusion principle* requires that there will be room for two electrons per unit cell of the lattice in an energy band. If each atom contributes one bonding electron, the valence band will be only half filled.

From the above it might be expected that a linear polymer backbone consisting of many strongly interacting co-planar p_z orbitals, each of which contributes one electron to the resultant continuous π-electron system, would behave as a one-dimensional metal with a half-filled conduction band. In chemical terms, this is a *conjugated* chain and may be represented by a system of alternating single and double bonds. It turns out that, for one-dimensional systems, such a chain can more efficiently lower its energy by introducing *bond alternation* (alternating short and long bonds). This limits the extent of electronic delocalization that can take place along the backbone. The effect is to open an energy gap in the electronic structure of the polymer. All conjugated polymers are large band gap semiconductors, with band gaps more than about 1.5 eV, rather than metals. Figure A.11(b) shows the example of *trans*-polyacetylene.

In charge-transfer materials, such as TCNQ (section 4.6), the molecules are planar and stack easily in the solid state. The highest molecular orbital of $TCNQ^-$ is a π orbital and a linear combination of atomic p_z orbitals is obtained, where z is the direction normal to the molecular plane. Interaction of these π molecular orbitals leads to the formation of a conduction band in the solid. It is along these overlapping π orbitals that electron motion occurs.

Further reading

Ferraro, J. R. and Williams, J. M. (1987) *Introduction to Synthetic Electrical Conductors*, Academic Press, Orlando

McMurray, J. (1984) *Organic Chemistry*, Brooks-Cole, Monterey

Pauling, L. (1960) *The Nature of the Chemical Bond*, 3rd edition, Cornell University Press, New York

Streitwieser, A., Jr. and Heathcock, C. H. (1976) *Introduction to Organic Chemistry*, Macmillan Publishing, New York

Appendix B

Interaction of electromagnetic radiation with organic thin films

B.1 Electromagnetic radiation

Besides what is commonly called light, *electromagnetic radiation* includes radiation of longer (infrared, microwave) and shorter (ultraviolet, X-ray) wavelengths (see p. xvii). As the name implies electromagnetic (EM) radiation contains both electric field **E** and magnetic field **B** components. The use of the bold typeface indicates that these are vector quantities. The relationship between the electric and magnetic fields is best illustrated by considering *plane-polarized* radiation. Here the electric vector is confined to a single plane. Figure B.1 depicts such radiation of wavelength λ travelling with *phase velocity* c (the velocity at which the crests of the wave travel) in a vacuum ($c = 2.998 \times 10^8 \, \text{m s}^{-1}$) along the x-axis. The electric component of the radiation is in the form of an oscillating electric field and the magnetic component is an oscillating magnetic field. These fields are orthogonal and are also at right angles to the direction of propagation of the radiation. The plane of polarization is conventionally taken to be the plane containing the direction of the electric field. *Unpolarized radiation, or radiation of an arbitrary polarization, can always be resolved into two orthogonally polarized waves.* If the two electric field components possess a constant phase difference and equal amplitudes, the resultant EM wave is said to be *circularly polarized*. If the amplitudes differ, then the wave is *elliptically polarized*. (Plane and circular polarizations are special cases of elliptical polarization.)

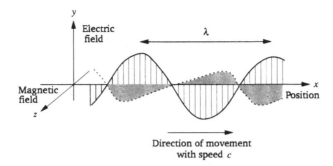

Figure B.1 Schematic representation of an electromagnetic wave. (λ – wavelength.)

In many experiments polarized EM radiation is used to probe thin layers deposited onto planar substrates. Two important measuring arrangements may be distinguished: *p-polarized* or *transverse-magnetic* (TM) incident radiation, in which the electric field vector is in the plane of incidence of the EM wave; and *s-polarized* or *transverse-electric* (TE) incident radiation, where the electric vector is perpendicular to the plane of incidence. These are contrasted in figure B.2.

B.2   Interaction with solids

A static electric field can induce a dipole moment in a material. In a similar way, the *oscillating electric field of an EM wave produces oscillating dipoles in matter*. In turn, these oscillating dipoles produce their own oscillating electric and magnetic fields which radiate EM waves. These combine with the incident wave to produce an electromagnetic wave that travels through the material with a velocity slower than the velocity of light in vacuum. If v is the phase velocity of light in the material and c is the velocity of light in vacuum, then

$$n = \frac{c}{v} \tag{B.1}$$

where n is called the *refractive index*. Typical values are 1.0003 for air, 1.33 for water and about 1.5 for glass. The refractive index may also be linked to the *relative permittivity* (or *dielectric constant*) ϵ_r of the medium

$$n = \sqrt{\epsilon_r} \tag{B.2}$$

Both refractive index and permittivity are complex quantities. For example

$$n = n' - jn'' \tag{B.3}$$

where the real part n' controls effects such as reflection, refraction (by *Snell's law*) and the velocity of propagation and the imaginary part n'' represents loss

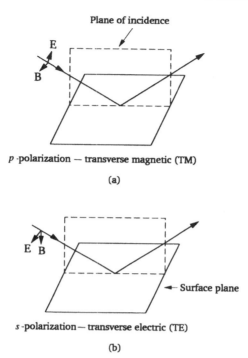

Figure B.2 *Diagrams showing s- and p-polarized EM waves incident on a surface. For p-polarized (TM) waves, the electric vector is in the plane of incidence (a). For s- (TE) polarization, the electric vector is perpendicular to the plane of incidence of the EM wave (b).*

(i.e., absorption in the material). Because light of various wavelengths produces different effects on the charges present in matter, the velocity at which light propagates through a material depends on the wavelength of the incident light. *This means that the refractive index and permittivity also depend on wavelength.*

B.3 Linear optics

The response of a material to the electric and magnetic components of an EM wave is determined by its permittivity and *permeability*. For most of the organic compounds discussed in this book, the relative permeability μ_r is equal to unity, so the interaction with EM waves is due solely to the dielectric properties.

The electric field **E** induces an *electric displacement* **D** in a medium such that

$$\mathbf{D} = \hat{\epsilon}_r \epsilon_0 \mathbf{E} \qquad (B.4)$$

where ϵ_0 is the permittivity of free space ($8.854 \times 10^{-12}\,\mathrm{F\,m^{-1}}$). The electric field and resulting displacement are both vector quantities, each with three components along three mutually orthogonal directions. This means the permittivity is a *second rank tensor* (formally written as $\hat{\epsilon}_r$) and its components may be represented by a 3×3 matrix array.

For any nonoptically active dielectric, it is possible to choose the x-, y- and z-axes so that the off-diagonal elements in the 3×3 permittivity tensor array are zero. This process is called *diagonalizing* the matrix and the resulting directions are called the *principal axes*. In matrix terms, equation (B.4) can then be written

$$\begin{pmatrix} D_x \\ D_y \\ D_z \end{pmatrix} = \epsilon_0 \begin{pmatrix} \epsilon_{11} & 0 & 0 \\ 0 & \epsilon_{22} & 0 \\ 0 & 0 & \epsilon_{33} \end{pmatrix} \begin{pmatrix} E_x \\ E_y \\ E_z \end{pmatrix} \tag{B.5}$$

i.e.,

$$D_x = \epsilon_0 \epsilon_{11} E_x$$
$$D_y = \epsilon_0 \epsilon_{22} E_y \tag{B.6}$$
$$D_z = \epsilon_0 \epsilon_{33} E_z$$

In an isotropic medium the induced polarization is independent of the electric field direction, so that $\epsilon_{11} = \epsilon_{22} = \epsilon_{33}$. Therefore, all propagation directions experience the same refractive index.

For an anisotropic medium the situation changes. In general there are two possible values of the phase velocity for a given direction of propagation. These are associated with the mutually orthogonal polarizations of the light waves. The two polarizations define the *ordinary* and *extraordinary* rays and possess distinct refractive indices, and therefore different angles of refraction at an interface. Hence, when light of an arbitrary polarization propagates through an anisotropic medium, it can be considered to consist of two independent waves, which travel with different velocities. Media exhibiting such effects are said to be *birefringent*. Two levels of anisotropy may be distinguished.

(i) *uniaxial medium*: $\epsilon_{11} = \epsilon_{22} \neq \epsilon_{33}$
   For most ray directions and polarizations, both ordinary (refractive index $= \sqrt{\epsilon_{11}} = \sqrt{\epsilon_{22}}$) and extraordinary rays (refractive index $= \sqrt{\epsilon_{33}}$) are generated. The exception is for light travelling in the z direction (the *optic axis*) where all polarizations are governed by the ordinary refractive index.

(ii) *biaxial medium*: $\epsilon_{11} \neq \epsilon_{22} \neq \epsilon_{33}$
   Here there are two optic axes of transmission along which the velocities of the two orthogonally polarized waves are the same. Otherwise ordinary and extraordinary rays are produced.

The molecules that are used to form an LB film are typically rod-like (chapter 2, figure 2.3). The films will possess a biaxial symmetry. However $\epsilon_{11} \approx \epsilon_{22}$ and the situation will approximate to case (i), with two refractive indices – along the rod and perpendicular to it.

B.4 Nonlinear optics

A second definition of electric displacement $\mathbf{D}$ to that provided by equation (B.4) is

$$\mathbf{D} = \epsilon_0 \mathbf{E} + \mathbf{P} \qquad (B.7)$$

where $\mathbf{P}$ is the *polarization* induced in the material by the electric field. Polarization is the induced charge per unit area or the dipole moment per unit volume; the units are $[\text{C m}^{-2}]$.

In *linear optics* the electric field is linearly related to the polarization, i.e.,

$$\mathbf{P} = \epsilon_0 \hat{\chi} \mathbf{E} \qquad (B.8)$$

where $\hat{\chi}$ $(= \hat{\epsilon}_{\text{r}} - 1)$ is the second rank *susceptibility* tensor. For *nonlinear optics* the above equation is replaced by a series expansion

$$\frac{\mathbf{P}}{\epsilon_0} = \hat{\chi}^{(1)} \mathbf{E} + \hat{\chi}^{(2)} \mathbf{E} \cdot \mathbf{E} + \hat{\chi}^{(3)} \mathbf{E} \cdot \mathbf{E} \cdot \mathbf{E} + \dots \qquad (B.9)$$

$\hat{\chi}^{(2)}$ is called the second-order susceptibility tensor (with SI units $[\text{C V}^{-2}]$) and $\hat{\chi}^{(3)}$ the third-order susceptibility tensor. The second-order term is responsible for effects such as *second-harmonic generation* and the *linear electrooptic effect (Pockels' effect)*, while the third-order term is responsible for *third-harmonic generation* and the *quadratic electrooptic effect (Kerr effect)*. An important consequence of crystal symmetry is that in a material possessing a *centre of inversion* (a *centrosymmetric* material), the second-order non-linear susceptibility $\hat{\chi}^{(2)}$ is zero.

Nonlinear properties are normally measured on macroscopic samples that consist of many individual molecules. On the microscopic level, a molecule placed in an electric field experiences a polarizing effect through a change in its dipole moment $\boldsymbol{\mu}$. Its relation to the electric field may also be expressed by a power series

$$\boldsymbol{\mu} = \hat{\alpha} \mathbf{E} + \hat{\beta} \mathbf{E} \cdot \mathbf{E} + \hat{\gamma} \mathbf{E} \cdot \mathbf{E} \cdot \mathbf{E} + \dots \qquad (B.10)$$

where $\hat{\alpha}$ is the linear polarizability and $\hat{\beta}$, $\hat{\gamma}$ etc. (all tensor quantities) are the higher-order polarizabilities, or *hyperpolarizabilities*. Centrosymmetric molecules will respond to the electric field of the optical wave to give equal and opposite polarization as the phase of the wave changes through $180°$ and therefore will have zero β coefficients. Large β coefficients (with SI

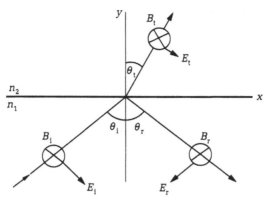

Figure B.3 Reflection and refraction of an electromagnetic wave at a boundary between two dielectric media. The incident wave is p-polarized, with its electric vector in the plane of incidence. ⊗ represents a magnetic vector perpendicular to the plane of the figure.

units [$C\,m^3\,V^{-2}$]) arise through the presence of an asymmetric mobile π-electron system and low lying charge-transfer states in molecules. Most organic molecules for second-order nonlinear optics are based on the construction: donor group/π-electron system/acceptor group.

B.5 Reflection and transmission from thin films

When electromagnetic waves are incident on the interface between two dielectrics, the familiar phenomena of reflection and refraction take place. Figure B.3 shows a simple case in which a *p*-polarized EM wave is incident, at an angle θ_i, to the boundary between two dielectric media of refractive indices n_1 and n_2. Usually there will be a reflected beam at an angle θ_r and a transmitted (refracted) beam at θ_t. Application of the boundary conditions that the normal component of **B** and the tangential component of **E** are continuous at the interface gives the following relations

$$\theta_i = \theta_r \tag{B.11}$$

$$\frac{\sin\theta_i}{\sin\theta_t} = \frac{n_2}{n_1} \tag{B.12}$$

Equations (B.11) and (B.12) are, of course, familiar as the *law of reflection* and *Snell's law of refraction*. (These two equations also hold for *s*-polarized incident radiation.)

The proportions of the incident electric field *amplitude* that are reflected may also be evaluated. For *p*-polarized radiation, the ratio E_r/E_i defines the *reflection coefficient* $r_{\|}$ *and* E_t/E_i *is the transmission coefficient* $t_{\|}$

$$r_{\|} = \frac{E_r}{E_i} = \frac{\tan(\theta_i - \theta_t)}{\tan(\theta_i + \theta_t)} \tag{B.13}$$

$$t_{\|} = \frac{E_t}{E_i} = \frac{2\cos\theta_i \sin\theta_t}{\sin(\theta_i + \theta_t)\cos(\theta_i - \theta_t)} \tag{B.14}$$

Equations (B.13) and (B.14) are known as *Fresnel* equations and apply at optical frequencies to transparent media where the refractive indices are real quantities (i.e., in equation (B.3), $n'' = 0$). Equivalent relationships, for $r_\perp$ and $t_\perp$, may be obtained for *s*-polarized incident radiation. The coefficients r and t also give the phases of the beams. If positive, there is no change in phase; if negative, the phase changes by π.

There is a particular angle of incidence at which $r_{\|}$ becomes zero while $r_\perp$ remains finite. This is called the *Brewster angle* θ_B and is given by

$$\tan\theta_B = \frac{n_2}{n_1} \tag{B.15}$$

For incidence from air to glass, ($n_1 = 1$; $n_2 = 1.5$) $n_2/n_1 \approx 1.5$ and θ_B is approximately $56°$. At this angle any incident light can only be reflected with **E** perpendicular to the plane of incidence, giving a method of producing linearly polarized light from an unpolarized beam.

The *power* (or intensity) of an EM wave reflected or transmitted from an interface is related to the *square of the electric field amplitude. Reflecting R or transmitting powers T* may be defined

$$R = r^2; \qquad T = t^2 \tag{B.16}$$

At normal incidence (i.e., $\theta_i = 0$), R and T are given for both states of polarization by

$$R = \left(\frac{n_1 - n_2}{n_1 + n_2}\right)^2 \tag{B.17}$$

$$T = 1 - R = \frac{4n_1 n_2}{(n_1 + n_2)^2} \tag{B.18}$$

Equation (B.17) predicts R is approximately 4% for radiation incident from air to a glass surface.

When radiation is incident from a dense to a less dense medium, Snell's law gives a *critical angle of incidence* θ_C at which there is 100% reflection (for both polarizations). The situation is referred to as *total internal reflection*.

Figure B.4 summarizes the main features for the reflection of *p*- and *s*-polarized EM radiation at a vacuum/dielectric interface ($n_1 = 1$; $n_2 = n$).

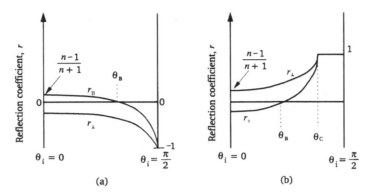

Figure B.4 Variation of reflected wave amplitude *with angle of incidence for: (a) waves incident from a vacuum (n = 1) to a dielectric with refractive index n; and (b) waves incident from inside the dielectric to a vacuum. ($r_{||}$ – reflection coefficient for p-polarized incident radiation; $r_{\perp}$ – reflection coefficient for s-polarized incident radiation. θ_B – Brewster angle; θ_C – critical angle.)*

If the EM radiation is incident from a transparent to an absorbing medium, then the real refractive index n_2 in equation (B.12) is replaced by $n_2' - jn_2''$, i.e.,

$$\sin \theta_t = \frac{n_1 \sin \theta_i}{n_2' - jn_2''} \tag{B.19}$$

Therefore, θ_t is complex and does not represent the angle of refraction, except for the special case $\theta_i = \theta_t = 0$. Here the Fresnel reflection coefficients (identical for both components of polarization) are given by

$$r_{||} = r_{\perp} = \frac{n_1 - n_2' + jn_2''}{n_1 + n_2' - jn_2''} \tag{B.20}$$

which gives the following expression for the reflecting power

$$R = \frac{(n_1 - n_2)^2 + (n_2'')^2}{(n_1 + n_2)^2 + (n_2'')^2} \tag{B.21}$$

The above ideas can now be applied to the situation in which there is more than one interface, for example that of a thin film on a solid substrate. Figure B.5 shows a schematic diagram for this. Radiation will first be reflected at the air/film boundary; the Fresnel equations for the appropriate polarization will govern the amounts of light transmitted and reflected. The transmitted radiation will undergo further reflection and transmission at the film/substrate interface. The overall air/film/substrate electric field amplitude reflection coefficient r is obtained by combining the Fresnel coefficients (for the appropriate polarization) after allowing for a change of phase β across

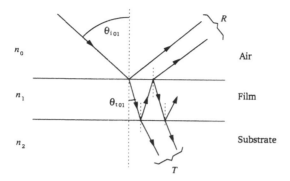

Figure B.5 Multiple transmission and reflection from an air/thin film/substrate combination.

the film (thickness d). The expression is given by

$$r = \frac{r_{01} + r_{12}\exp(-\text{j}2\beta)}{1 + r_{01}r_{12}\exp(-\text{j}2\beta)} \tag{B.22}$$

where r_{01}, r_{12} are the Fresnel coefficients for the air/film and film/substrate boundaries, respectively. The phase change β across the film of thickness d is

$$\beta = 2\pi\left(\frac{d}{\lambda}\right)n_1\cos\theta_{t01} = 2\pi\left(\frac{d}{\lambda}\right)(n_1^2 - n_0^2\sin^2\theta_{i01})^{0.5} \tag{B.23}$$

The transmitted amplitude t is given by

$$t = \frac{t_{01}t_{12}\exp(-\text{j}\beta)}{1 + r_{01}r_{12}\exp(-2\text{j}\beta)} \tag{B.24}$$

Equations (B.22) and (B.24) are generally valid. For nonnormal incidence each takes two possible forms, depending on the state of polarization of the incident EM radiation. If the film is absorbing, or if it is bound by absorbing media, then the values of n_0, n_1, n_2 are replaced by the corresponding complex quantities. The resulting equations for r and t become somewhat cumbersome, although readily calculable using a computer.

The oscillatory nature of equations (B.22) and (B.24) is a result of the constructive and destructive interference from the multiple reflected and transmitted EM waves. By appropriate choice of the thin film thickness ($d = \lambda/4$) and refractive index ($n_1 = \sqrt{n_2}$), it is possible to obtain zero reflectance at normal incidence for a particular wavelength (the principle of an antireflection coating).

Under certain conditions the equations for the transmitted and reflected intensities (powers) from a thin absorbing film/transparent substrate combination may be considerably simplified. For *normal incidence* and if *multiple reflections and interference can be neglected* (which will be the case if the film is sufficiently absorbing)

$$T = (1 - R)^2 \exp\left(\frac{-4\pi n''}{\lambda}\right) \tag{B.25}$$

Together with equation (B.21), the above relation provides a convenient method of obtaining n' and n'' from measurements of T and R on the same film.

B.6 Absorption processes

The *intensity* of light passing through an absorbing material is reduced according to *Beer's law*

$$I = I_0 \, e^{-\alpha l} \tag{B.26}$$

where I is the measured intensity after passing through the material, I_0 is the initial intensity, α is the *absorption coefficient* and l is the path length. This is often written in a more convenient form as

$$A = -\log_{10}\left(\frac{I_0}{I}\right) \tag{B.27}$$

where A is the *absorbance* of the sample and is given by $A = (\log_{10} e)\alpha l = 0.43\alpha l$. The absorption coefficient is related to the imaginary part of the complex refractive index by

$$\alpha = \frac{4\pi n''}{\lambda} \tag{B.28}$$

where λ is the wavelength of the light.

The *Beer–Lambert law* introduces the concentration of an absorbing species into the above relationship and is used in work with solutions. The absorbance of a sample can be expressed

$$bcl = -\log_{10}\left(\frac{I_0}{I}\right) = A \tag{B.29}$$

where b is the molar absorption coefficient [litre mol^{-1} cm^{-1}], c the molar concentration of the solution [mol litre^{-1}] or [M] and l is the path length in centimetres. The value of b is usually quoted at an absorption maximum. Since the absorbance is proportional to the concentration of the solution, a plot of A versus c should yield a straight line. This relationship is adhered to in many materials. However, at high concentrations, changes in the position or intensity of the absorption maximum, caused by the formation of molecular aggregates such as *dimers* or *trimers*, often result in a deviation from the Beer–Lambert law. The interaction between the dye molecules and the solvent (*solvatochromism*) can also give deviations from this law.

In terms of electronic energy levels, a molecule in its ground state can absorb a photon of light if the photon energy is equal to the difference between the two energy levels in the system. When this occurs, the molecule is excited into a higher energy state so that

$$E_1 - E_0 = h\nu \qquad\qquad (B.30)$$

where h is Planck's constant (6.626×10^{-34} J s), ν is the frequency of the radiation in [Hz], and E_1 and E_0 are the excited and ground states of the molecule, respectively. The minimum photon energy required for absorption will be when E_0 corresponds to the *highest occupied molecular orbital* (HOMO) and E_1 to the *lowest unoccupied molecular orbital* (LUMO).

Absorption of light may produce excitation of an electron from a bonding π to a corresponding antibonding molecular orbit (appendix A). The latter is called a π^* orbital and the absorption is referred to as a $\pi \to \pi^*$ transition. (This is analogous to the transfer of an electron from the valence band to the conduction band in a semiconductor.) The π molecular orbitals of conjugated systems extend over several atoms and there are as many π levels as there are conjugated bonds. As the length of conjugation increases so does the wavelength of the absorption band.

Nonbonding n electrons form an inner or *lone pair* of electrons. These are normally not held as tightly as σ electrons and they can also be excited to π^* molecular orbitals. This process results in a so-called $n \to \pi^*$ transition and usually occurs at a longer wavelength (lower energy) than the $\pi \to \pi^*$ transition. Absorption due to $\sigma \to \sigma^*$ transitions is usually at a short wavelength (in the ultraviolet). The various electronic energy levels associated with both simple molecules (discrete levels) and complex unsaturated molecules (bands of energies) are shown in figure B.6.

The promotion of an electron from a lower energy level to a higher energy level does not produce a permanent dipole moment. However, at some intermediate transient stage in the transition, a lack of symmetry in the electron density distribution exists. The *transition dipole moment* (having direction as well as magnitude – section 2.2) that results enables light absorption to take place. Light will not be absorbed completely unless the oscillating electric field is parallel to the transition moment. This dipole moment is fixed relative to the molecular structure. In ethylene, the $\pi \to \pi^*$ transition, which is associated with the C=C bond, has its transition moment polarized along the bond, as shown in figure B.7. Complex molecules may have more than one transition moment.

For dye molecules on a solid plate, the orientation (if any) can be conveniently investigated using polarized light. For instance, if a wavelength at which

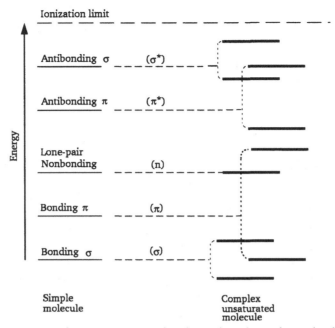

Figure B.6 Bond energies associated with simple and complex molecules.

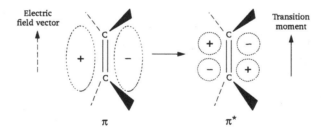

Figure B.7 Orientation of the transition dipole moment in ethylene.

a molecule absorbs is chosen, then direct comparison of the absorption intensities of the *s*- and *p*-radiation (at the same angle of incidence) enables the
average orientation of the transition moments to be determined.

Most organic molecules have an even number of electrons, with all electrons paired. Within each pair, the opposing electron spins cancel and the
molecule has no net electronic spin. Such an electronic structure is called a
singlet state. When a ground state singlet absorbs a photon of sufficient
energy, it is converted to an excited singlet state in which the spin of the
excited electron is not altered. The process is so fast that the excited state
has the *same geometry of bond distances and bond angles*. This is the
Franck–Condon principle. However, the most stable geometry of the excited

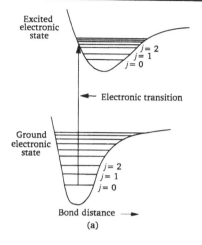

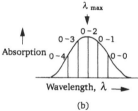

Figure B.8(a) Morse curves for the ground and excited states of a molecule.
(b) The origin of the shape of absorption bands.

state often differs from that of the ground state so that the excited electronic
state is often formed in an excited *vibrational* state.

Shapes of absorption bands can be explained by reference to *Morse curves*.
These are plots of potential energy as a function of the bond distance. In figure
B.8(a), Morse curves corresponding to a ground and an electronically-excited
state of a particular molecule are shown. The horizontal lines represent the
vibrational levels of the electronic states and each of these has a vibrational
quantum number associated with it, $j = 0, 1, 2, 3, \ldots, n$. *The excited state
is represented by a Morse curve which is displaced vertically (higher
energy) and horizontally (due to vibration increased bond length) from
that of the ground state.* By the Franck–Condon principle, electronic
transitions are represented on the Morse curve by vertical lines.

The probability and intensity of an electronic transition will not be a
maximum between the $j = 0$ levels of the ground and excited states. Because
of the relative displacement of the two Morse curves, the maximum absorption will occur between the $j = 0$ level of the ground state and a higher energy
level in the excited state (for instance $j = 2$ in figure B.8(a)). This corresponds
to the wavelength for maximum absorption, λ_{max}. On each side of this, the

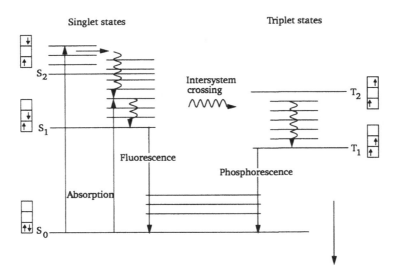

Figure B.9 Schematic diagram showing the radiative and nonradiative decay of an excited molecule. The orientations of the electron spins are shown in the boxes next to each state.

absorption intensities will decrease to zero, producing the familiar bell-shaped absorption band shown in figure B.8(b). In polyatomic molecules the Morse curves are replaced by polydimensional surfaces and the number of allowed transitions will become very large. Consequently the absorption bands become smooth curves. However, sometimes vibrational fine structure can be seen.

B.7 Emission processes

Luminescence is a term that is used to describe the emission of light by a substance caused by any process other than a rise in temperature. Molecules may emit a photon of light when they decay from an electronically excited state to the ground state. The excitation leading to this emission may be caused by a photon (*photoluminescence*), an electron (*electroluminescence*) or a chemical reaction (*chemiluminescence*).

The processes leading to emission can be considered in terms of the energy levels in the molecule. Figure B.9 (a *Jablonski diagram*) shows a very simplified energy diagram illustrating the various processes that can occur. Following promotion to a vibrationally excited singlet state (S_1 or S_2 in the figure), the molecule usually relaxes to the lowest vibrational level of S_1, the energy being lost as heat via intermolecular collisions. This process takes about 10^{-11} s, or approximately 10^2 vibrations of the molecule. The lifetime of S_1 in its lowest vibrational state is longer, 10^{-8} to 10^{-7} s. This state may then decay to the ground state with the emission of a photon. This emission, which occurs at a longer wavelength than absorption, is termed *fluorescence* (compare the energy of absorption for $j = 0$ in the ground state to $j = 2$ in the excited state in figure B.8(a) to that of emission, $j = 0$ in the excited state to $j = 0$ in the ground state).

Alternatively, the energy may be transferred by *intersystem crossing* to a *triplet state*. In such a state the spins of the two electrons are parallel and a transition to the ground state, with the emission of a photon, involves a change of spin. This is a *spin-forbidden* transition and triplet states are fairly long lived, with lifetimes of greater than 10^{-5} s. The triplet state can decay to S_0, emitting a photon. This process is known as *phosphorescence* and may persist for seconds or even longer after the incident excitation has ceased. Nonradiative transitions from the excited singlet and triplet states to the ground state are also possible. In these instances the original light quantum is converted into heat.

Further reading

Chopra, K. L. (1969) *Thin Film Phenomena*, McGraw-Hill, New York

Dobbs, E. R. (1985) *Electromagnetic Waves*, Routledge & Kegan Paul, London

Duffin, W. J. (1990) *Electricity and Magnetism*, 4th edition, McGraw-Hill, London

Heavens, O. S. (1970) *Thin Film Physics*, Methuen, London

Heavens, O. S. (1965) *Optical Properties of Thin Films*, Dover, New York

Nye, J. F. (1985) *Physical Properties of Crystals*, Oxford University Press, Oxford

Prasad, P. N. and Williams, D. J. (1991) *Introduction to Nonlinear Optical Effects in Molecules and Polymers*, Wiley-Interscience, New York

Wilson, J. and Hawkes, J. F. B. (1983) *Optoelectronics: An Introduction*, Prentice-Hall, New Jersey

Yariv, A. (1991) *Optical Electronics*, Saunders College Pub., Philadelphia

Appendix C

Crystallography

C.1 The crystal lattice

An ideal crystal contains atoms arranged in a repetitive three-dimensional pattern. If each repeat unit of this pattern, which may be an atom or group of atoms, is taken as a point then a three-dimensional *point lattice* is created. A *space lattice*, such as that shown in figure C.1, is obtained when lines are drawn connecting the points of the point lattice. The space lattice is composed of box-like units, the dimensions of which are fixed by the distances between the points in the three noncoplanar directions x, y and z. These are known as *unit cells* and the crystal structure has a periodicity (based on the contents of these cells) represented by the translation of the original unit of pattern along the three directions x, y and z. These directions are called the *crystallographic axes*. Any directions may, in principle, be chosen as the crystallographic axes. However, it is useful to select a set of axes which bears a close resemblance to the symmetry of the crystal. This can result in x, y and z directions that are *not at right angles to one another*. In figure C.1, the angle between the y and z axes is designated α, between the z and x axes, β, and between the x and y axes, γ. The measured edge lengths of the unit cell along the x, y and z axes are commonly given the symbols a_0, b_0 and c_0.

C.2 Crystal systems

As noted above, the unit cells of which a space lattice are composed do not necessarily have their three axes at right angles. The lengths of the sides can also vary from the case where they are all equal to the case where no two of them are the same. Crystals can belong to seven possible *crystal systems*, characterized by the geometry of the unit cell.

Table C.1 *The seven crystal systems.*

Crystal system	Axes	Angles
Cubic	$a_0 = b_0 = c_0$	$\alpha = \beta = \gamma = 90°$
Tetragonal	$a_0 = b_0 \neq c_0$	$\alpha = \beta = \gamma = 90°$
Orthorhombic	$a_0 \neq b_0 \neq c_0$	$\alpha = \beta = \gamma = 90°$
Monoclinic	$a_0 \neq b_0 \neq c_0$	$\alpha = \gamma = 90° \neq \beta$
Hexagonal	$a_0 = b_0 \neq c_0$	$\alpha = \beta = 90°; \gamma = 120°$
Rhombohedral	$a_0 = b_0 = c_0$	$\alpha = \beta = \gamma \neq 90°: <120°$
Triclinic	$a_0 \neq b_0 \neq c_0$	$\alpha \neq \beta \neq \gamma \neq 90°$

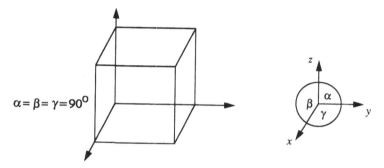

Figure C.1 *A space lattice showing a crystallographic unit cell of dimensions* a_0, b_0 *and* c_0. *The angles between the axes are designated as* α, β *and* γ.

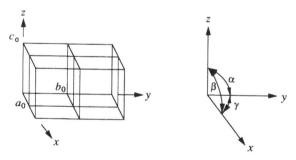

Figure C.2 *Four unit cells, each of dimensions* $a_0 \times b_0 \times c_0$.

Figure C.2 shows a group of four typical unit cells with a right-handed set of axes; the angles between the axes, α, β and γ, are, therefore, all equal to 90°. The relationships between the cell dimensions, a_0, b_0 and c_0, and the angles between the crystallographic axis for the seven possible crystal systems are given in table C.1.

In the simplest lattices based on the crystal systems shown in table C.1 (*primitive* unit cells), the lattice points are positioned at the corners of the cell. However, the monoclinic, orthorhombic, tetragonal and cubic systems can

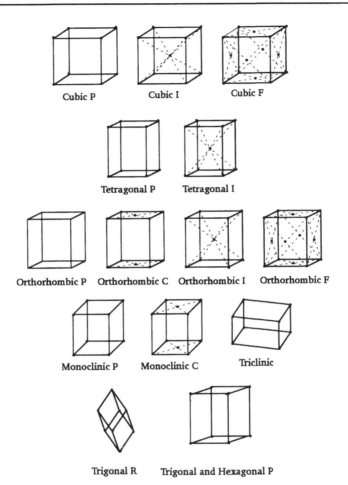

Figure C.3 Unit cells of the 14 Bravais lattices. (P – primitive unit cell; C – base-centred unit cell; F – face-centred unit cell; I – body-centred unit cell.)

also have cells that possess additional lattice points. These can occur at the centre of faces or in the middle of the body diagonal, leading to *base-centred* (C), *face-centred* (F) and *body-centred* (I) unit cells. There are seven of these lattices and with the seven primitive (P) lattices they constitute the fourteen distinct *Bravais lattices* shown in figure C.3. Whereas a primitive cell possesses one lattice point per unit cell (each of the eight corners being shared by eight cells), body-centred and base-centred cells possess two points and face-centred cells four points.

C.3 Miller indices

Many sets of planes can be drawn through the lattice points of a crystal structure and diffraction of X-rays by the crystal can be treated as reflections of the

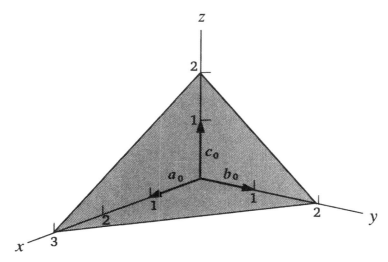

Figure C.4 (233) plane

X-ray beam by these planes. It is desirable, therefore, to be able to describe each set of planes uniquely. This is accomplished using *Miller indices*, which were originally derived to describe crystal faces, but can be applied equally well to any plane or set of planes in a crystal. Miller indices are allocated to a plane in a crystal by first specifying its intercepts on the three crystal axes, in terms of the lattice constants. The reciprocals of these numbers are then taken and reduced to three integers, usually the smallest three integers, having the same ratio. The result is then enclosed in parenthesis: (hkl). For example, figure C.4 shows a crystal plane intercepting the x, y, z axes at $3a_0$, $2b_0$, $2c_0$. The reciprocals of these numbers are $\frac{1}{3}$, $\frac{1}{2}$, $\frac{1}{2}$. The smallest three integers having the same ratio are 2, 3, 3. Therefore, the Miller indices of the plane are (233). If one or more of the intercepts are at infinity (i.e., for planes parallel to a crystallographic axis), then the corresponding index is zero. Figure C.5 shows some important planes in a cubic crystal. Other conventions that are used include:

- $(\bar{h}\bar{k}\bar{l})$: for a plane that intercepts the x-axis on the negative side of the origin.
- $\{hkl\}$: for planes of equivalent symmetry, e.g., $\{100\}$ for (100), (010), (001), ($\bar{1}$00), (0$\bar{1}$0) and (00$\bar{1}$) in a cubic crystal.
- $[hkl]$: for a direction in a crystal, e.g., [100] is the x-axis in a cubic crystal.
- $\langle hkl \rangle$: for a full set of equivalent directions.

C.4   Distance between crystal planes

In crystallography, it is often necessary to calculate the perpendicular distances between successive planes in a series of planes (hkl). This is called

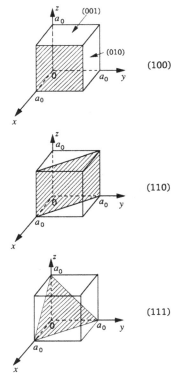

Figure C.5 Miller indices of some important planes in a cubic crystal (lattice constant – a_0).

the d_{hkl} spacing. For a cubic crystal system, this is easily obtained from simple geometry

$$d_{hkl} = \frac{a_0}{(h^2 + k^2 + l^2)^{1/2}} \tag{C.1}$$

The equivalent relationship for the orthorhombic systems is

$$d_{hkl} = \frac{1}{[(h/a_0)^2 + (k/b_0)^2 + (l/c_0)^2]^{1/2}} \tag{C.2}$$

Further reading

Phillips, F. C. (1971) *An Introduction to Crystallography*, Oliver and Boyd, Edinburgh
Milburn, G. H. W. (1973) *X-Ray Crystallography*, Butterworths, London

Materials index

Subject index

absorbance 179, 216
absorption coefficient 216
acceptor *see* electron donor and acceptor
accumulation 159
activation energy (for conduction) 132, 151
adsorbed layer 60–2
aggregate 180–2
  H– 181
  J– 181
aliphatic compound 69
alkyl chain 65, 68–9
alternate-layer film 40, 148, 150, 189
amorphous solid 1
amphiphilic molecule 15
anisotropic medium 2
antibonding *see* molecular orbital
anti conformation 200
area per molecule 17
aromatic compound 71
atomic force microscopy (AFM) 121–4
Auger electron spectroscopy (AES) 124

band gap *see* energy gap
band theory 203
bathochromic shift 181
Beer–Lambert law 216
Beer's law 216
benzene ring 70–1
  ortho-, meta-, para-positions 70
β coefficient in nonlinear optics 186, 211
biaxial medium 210
birefringence 210
black lipid membrane 48
Boltzmann statistics 33
bond alternation 205
bond length 202
bond strength 202
bonding *see* molecular orbital

Bragg's law 98
Bravais lattice 224
Brewster angle 213
  microscopy 23
Brillouin scattering 120

capacitance 42
  differential 158
centre of inversion 211
centrosymmetric 211
Cerenkov radiation *see* nonlinear optics
charge carrier 131
  density 131
  drift velocity 131
  mobility 131
chemiluminescence *see* luminescence
chiral molecule 4, 87
chromophore 73–4, 182–3
  orientation 182–3
cis-bond 69, 74, 95, 201
collapse of monolayer 18
component 9
compressibility 30
compression system 49–52
condensed phase *see* monolayer phase
conductance (electrical) 146–7
conduction band 203
conductivity *see* electrical conductivity
configuration *see* isomer configurational
conformation *see* isomer conformational
conjugated chain 205
correlation radius 101
critical angle 170, 213
critical micelle concentration 19
critical point 6
crystal 222
  axis 222
  crystalline solid 1

229

Printed in the United States
By Bookmasters